REAL RANDOM NUMBERS

Discovery, Generation & Encryption

REAL RANDOM NUMBERS

Discovery, Generation & Encryption

Professor N. C. Das

Harshit Ranjan

SHROFF PUBLISHERS & DISTRIBUTORS PVT. LTD.

Navi Mumbai

Real Random Numbers

Discovery, Generation & Encryption

By Professor. N. C. Das & Harshit Ranjan

First International Print: June 2026

ISBN: 978-93-6808-093-0

Printed at SAP Print Solutions Pvt. Ltd., Mumbai. • Published by Shroff Publishers and Distributors Pvt. Ltd. B-103, Railway Commercial Complex, Sector 3, Sanpada (E), Navi Mumbai 400705 • Tel: (91 22) 4158 4158 WhatsApp +91 7304487700 • E-mail: spdorders@shroffpublishers.com • URL: www.shroffpublishers.com

Dedicated to

1. The small simple monograph that bears elegant look but unearths incredible search is being dedicated to all its potential users who have been desirous for Real Random Numbers at least since about a century for optimally safe and sound message encryption or for any other appropriate application.
2. The book so written under painful sorrow of loosing loving life partner of Senior Author's bearing her name now Late Lakhi Rani Das is being dedicated to herself along with all her future generations.

Such dedication is being made to recollect invaluable contributions of those, who have made/done for the production and publication of the book. Lest, they be forgotten in time and space.

Monograph

Its semi-poetic verse,

"This simple and short monograph is so nice,

that it has melted the knowledge gap which remained iced, for about a century.

Every one may like to read it Twice and Thrice, till he gets Expertise.

It remains user friendly even it quotes, refers and uses, higher mathematics.

"It makes users to travel historic path, right from Kapil to Hilbert via Bhaskaracharya.

It reveals the secret that remained concealed even to Knuth's exhaustive references and tc Hromkroic, Stallings and Schneier".

However, Agnew happened to be authors' torch bearer, whose writings made him to go up to Kapil, tc find congruence to Random.

But Kapil's such similitude was his aspired virtue in Saintly Person, not for any Number, whatsoever

On the other hand a long historic journey from Bhaskaracharya to Hilbert for numbers, but not for the same to be of random.

It is authors' meticulous search that such REAL RANDOM NUMBERS, have been discovered.

Thus it is, evolutionary discovery for the same, credit-worthy is author alone.

Random is Random, it does not need to be certified and/or testified that it is indeed the 'REAI RANDOM NUMBER' (RRNs).

- Professor N. C. Das
Harshit Ranjar

Preamble

As you read, you know that 'you! didn't knew ';

And, that you are still inquisitive to read; further, and furthermore.

When you have reached that state; what does it mean?

It means "you have understood, what you have read."

Where, does your understanding take you?

You are at decisive state either to accept or not to accept.

If accepted, you wish to develop your understanding further.

Now, you have reached at progressive state, wherein you would be able to put your understanding to action.

It is here or even before that 'Random choice' puts you often on a right-track, in order to proceed for 'Robust Decision' directed towards 'Optimal Action'.

- Professor N. C. Das

Preface

Myself an author along with all others have to asses where we are at this critical juncture, where elites of whole World are acclaiming for AI (The Artificial Intelligence) and who or what happens to be its most powerful player.

Every creature is required to act. Action essentially requires decision and the decision, requires relevant prior knowledge. As we often may not know the future, yet decision followed by action is absolute necessity. Gauzing the adversity of situation the decision must be robust. Nothing can help you more than the choice of RANDOM and possibly getting its instantaneous repetitions. It is here the most powerful player that could be identified by leaders of the recent relevant past.

A book on RANDOM is perhaps the first venture on the topic specially that on Real Random which hitherto has been considered virtually impossible to be obtained. But it is my mind that threw the responsibility to my soul to take such an arduous task.

As expressed herein above it was considered most difficult to get the same in its REAL form. How difficult that is can be known as the Chapter advances, from Chapters 1 to 3.

Further it is to be expressed that of seventy seven years my acquaintance with theory of statistics, probability and/or their applications : it has not been seen a book on such subjects, which do not use the term 'random', at least once. Incidentally, to my utter surprise, a Sanskrit couplet was found which was due to Maharishi KAPIL, written millennium years before, in his *SHAKHYA-KARIKA*, the prime text of *SHAKHYA* philosophy. From there we quote here only its two terms *'aakhyayika-vitrahitah* and *'parvaad-vi-var-jitash-cha-pi'* these are enough to imply that he(the KAPIL) then itself aspired celebrant person to posses the virtues of these two terms. That made his concept, similitude of 'real random', the term being sought after and to be brought to light in this text. That has been done in very first Chapter , in Chapter 9 and elsewhere also.

In prospecting paradigm the term randomization along with its origin and path-way, the 'Random Numbers' have become necessary tools for scientific methods, keeping in abeyance philosophical paradox of its prohibitive usage.

Future is unknown, but one has to act: action needs decision, decision must be robust. An approach for robustness requires 'random number' and/ or 'randomization' to be the choicest.

What such 'random' yields?

The answer is, come it may whatever, we must have device to sustain and proceed ahead. The scientific method, we say, better the 'empirical method' so far, have not been able to find any other device, better than this (i.e, the Random). But such an assurance is only for Real-Random, not for its mockeries whatsoever. The term 'random' alone, or as an adverb may have to be taken as 'random- choice' or 'random- selection'.

Several computer-oriented, mathematically-based and statistically-tested attempts have been made and/or are being advanced but all for Pseudo-Random Generators. As it is the persisting wide-spread but doubtful hunch that REAL-RANDOM could not and cannot be found. Leading us to think, all such users are in 'hornet nest' implying to be in situation full of difficulties. That puts the whole WORLD in awful state, because of its ubiquitous usage, being wrongfully so rampant. Continuance of such action for scientific purposes is just like running a dancing puppet to imitate the celebrated dancer.

Facing those, the proposed text has to find 'solution' and that to be 'short' and 'simple'. The task is indeed very challenging, how to meet them is the moot question. A sincere attempt has to be made for at least to erect viable alternative to existing practice, better say malpractice in face of existing philosophical paradox.

The primary way to 'shorten' is to consider only the generation of 'Random Numbers', rather its 'real' form for the very foundation and crux of the problem. Thus to escape its several applicative derivatives like 'Randomized events', 'Randomization of variables', 'Random walk', 'Random Processes' and say 'Randomized Algorithms', as they all involve additional requirement, that is the term 'probability'. It is for such reasons almost everyone concerned is required to raise, citadels on weaker foundation. That could be naturally 'simple' if we may be out of the twin crusts of, 'modular arithmetic' and of 'large prime numbers' followed impliedly by several suggested statistical tests: as both the formers, form unbreakable inter-twin with each other, like chromosomal helix.

Also in sentence herein above, we have gotten several facets of the term 'random' and/or its adjectives/ adverbs, hence the term must find its interpretation and/or usage to be in accordance with contextual settings. Never the less, we must set up the undernoted moot question and its relevant answer.

Which is: Does True Random exist?

Though the Question is like asking 'Does GOD' exist? .

For knowing its answer let us at first refer KENO-UPANISHAT ||3||.

न तत्र चक्षुर्गच्छति न वाग्गच्छति नो मनः ।

न विद्मो न विजानीमो यथैतदनुशिष्यात् ॥ ३॥

na tatra cakṣurgacchati na vāggacchati no manaḥ .

na vidmo na vijānīmo yathaitadanuśiṣyāt .. 3..

The eye does not go there, nor speech, nor the mind. We do not know ; ' how to teach one about That?'.

The answer to the existence of both is only a word: that is 'Verily'.

But we have to search both, as They (both) are canopied, hence philosophically well encrypted.

Such question is not new, as for its answer, an Indian who may perhaps be knowing. That, we may quote 'Nasdiya Suctam' in Rig-VEDA 10-129-6 & 7, wherein HIS existence could at first, like present statistical test started with Null Hypothesis. In fact the entire *Nasdiya suktas* of seven stanzas could form such Null Hypotheses but as symbolic sample we have taken only one of those which is, 'about His non-existence'. Thus made probable by utterance:

इयं विसृंष्टिर्यतं आबभूव यदिं वा दधे यदिं वा न।

यो अस्याध्यंक्षः परमे व्योंमन्त्सो अङ्ग वेंद यदिं वा न वेदं॥ ७॥

Iyam visrushtir-yata aabhabhuuva yadi vaa dadhe yadi va na |

yo asyaadhyakshah parame vy-o-mantso anga ve-eda yadi va ne veda || 7 ||

Where did creation have its origin? Who is One that created it or did the One not create it?

That One alone perceives all from above and knows the beginning or may be even He doesn't?

As an answer to such doubt, aroused through an equivalent to above raised Null Hypothesis: an erudite Vedic scholar Madhuchhand-das (believably son of Vishva-mitra as mentioned in Rig-Veda) asserted:

सूर्यचन्द्रमसो धाता यथापूर्वमकल्पयत् । दिवं च पृथिवीं च अन्तरिक्षमथो स्वः ॥

sūrya-candramaso dhātā yathā-pūrvam akalpayat | divam ca pṛthivīṁ ca antarikṣam atho svaḥ ||

Literal Meaning:

"The Creator (Dhātā) re-formed (akalpayat) the Sun and the Moon (Sūrya and Candrama) as before (yathā-pūrvam); likewise, He formed the Heaven (divam), the Earth (pṛthivīm), the mid-space (antarikṣam), and the realm of Svarga (svah)."

Repeating the same once again:

"Surya-chandramoso dhata yatha-purvam kalp-yata:

Dwing-cha prith-wing cha-anterichh-matoo swah.

Import: Dhatar the great Creator, then formed in due order Sun and Moon, again He formed in order, Heaven and Earth, the regions of the air, and light.

His premises here, were based on the logic "if there is observable, perceivable and/or realizable 'Effect', there must be the 'Cause', that could be enormously powerful: to be called Dhatar (*Brahma*), the great Creator.

Thereafter Narayana, the other Vedic-saint independently joined him with his famous *Purush Suktam:* connoting the same as 'The Universal Soul, implying *PARMATMA*'.

Thus existence of God could usher-in, by the utterance of 'Om'. For the author here 'OM' means 'O! My Mother'.

Those were considered enough to reject the then Null-hypothesis, got raised herein above by *Nasdiya Suctam:* through which doubt was raised about HIS very existence.

Such a practice of statistics, has instructed that if any entity or thought cannot/ must not be found by objectively direct search, then one must find it, which he is required to reject for his direct search. That is what one gets by rejection of 'Null Hypothesis'.

Such a concept emerges from Badrayan's (1998, 2003), where his Brahmasutra enunciated :

"भवे-च उपलभ्धे" (bhave-ca upalabdhe)

"न-भवो ऊन्-उ-उ-पत्लभधे" (na-bhavo unu-upa-labhdhe)

'Bhave- cha Upalabdhe' and 'Na-bhavo Unu-upa-labhadhe'.

With their respective Imports:

If there is any perceptible/ realizable 'Effect' there must be 'Cause': in case there is no such 'Effect', no 'Cause' can be apprehended even suspected, much less can even be 'Hypothesized'. We must have sensitive device to measure the effective changes. That like one being excerpted.

"Relatively recently *Newton(1642-1727)* standing on the shoulders of Pierre *Gassendi(1592-1655)* an atomist*(Particle Physicist)* declared(1675) that Light was made of tiny particles (corpuscles),implying discrete, moving in straight lines. Refer: Whyntie & Pugh (2013). Why so?, Who causes so?, and Why He causes so?: also where are the causes for such effects, yet remain to be answered.

Thus the term 'Verily' could be established for the existence of God, but the same remains to be got established for the term 'Real-Random', by us the authors. That is the task to be got fulfilled by this monographic text.

In course of writing we floated little away : however returning back to the topic concerned, we have to keep in mind, though the requirement of 'Random' is for very large mass of people, including its huge communication world market: educational effort made so far for such a demanding mass or market has been very scanty. That has been realization felt by Hromkovic (2005).

Besides what is necessary is to gauze its appropriateness, meaning 'who needs what?'. To be more clear 'of what digital-length' such random numbers are needed by whom, at what price and how many are they. Also, what is the frequency of such demand? Such information is required, to design multi-optimal digital-lengths of such random numbers, needed for plethoric uses.

Through this small and simple monograph, attempt has been made, to search for Real Random Numbers, in order to settle about a century old philosophical paradox and then to provide a viable alternative to the existing practice, made operational only through Pseudo Random Numbers, which is only unethical mimicry of Real Random Numbers. The same is like the best known from underlying expression of Democritis quoted by Hromkovic's(op cit):

"A dwarf will be a dwarf even stands on the top mountain,

a giant will stay a giant even standing at the bottom of the deepest valley."

Full Knowledge, is seldom achievable, our recourse then is 'Real Random'.

The Real- Random is so robust that it can sustain you most odds: alas the same is also not a perfect device for all ills you may encounter in your life.

How and Where from, one may get Real Random Number is the objective hence the subject matter of this Monographic text. Even if partial success in such direction: the Text may claim it to be the Masterpiece work of importance.

It is at the End of the text we shall be able gauze whether the term 'Verily' could be established for its proposition or yet remains in some doubt. That still is mere Hypothesis.

Thus once again to summarize we may say " our aim is only to search viable alternative of 'Pseudo Random Number', not its substitute. Its substitution is the choice of the users. That is left to conscience, confidence, convenience and relative cost for changing occasions, on its usages by them on degree of confidentiality verses unauthorized decoding/ distortion or theft of the message.

The proposed text has to have two parts, virtually three parts but that of Appendix is not considered as a cogent part.

Its PART 1 on PRELIMINARIES, PHLOSOPICAL PARADOX and PATH-WAY to REAL RANDOM NUMBERS consisting first four chapters 1 to 4, where as its PART 2 on DISCOVERY , GENERATION and ENCRYPTION contains six chapters 5 to 10.

Chapter 1 begins with the necessity of random, thereby in its successive sections 2 to 11, dealing respectively with Introductory Remarks , its meaning, of facing the same, existing state, the spate of objections being faced, raising the question of safety, introducing Pseudo Random Numbers (PRNs), Operating conflicts , attempted definition for Real Random Numbers (RRNs) and ends with existence of its discrete congruence/ similitude.

Chapter 2 with its eight sections, is meant for quoting and/or referring some, renowned peers of their stands and views about persisting philosophical paradox, yet continuing for the use of PRNs.

Chapter 3 having ten sections is primarily meant for going in depth about the stand taken by such peers pertaining to related interdisciplinary fields like number theorists, probabilistic theory apart from experts of the field of cryptography.

Thus such in depth review, author got the clue for generating/ obtaining real random numbers (RRNs) from Agnew's (1988) exploratory investigations.

Chapter 4 has to have seventeen smaller sections on incredible search for much awaited RRNs the Real Random Numbers. Its section starts with Introductory Note then to 'The search begins' ,thereafter 'How and where from to get them (RRNs)?,followed by Author's Hype and Its(RRNs) Clue, here in Chapter 4. For further smaller section their contents have to be referred. That finishes PART 1 of the book.

PART 2 of the book begins with Chapter 5 and runs up to Chapter 10. Such chapters are of real importance for those who are desirous for several variants of RRNs. Fulfilling the main purpose for which the book is being written. We have proposed more than five methods those originating from Chapter 4 of Part 1 itself. Thereafter it runs from Chapters 5 to 9. In none of them there is any theorem but only a step or two of formulating algorithm followed by data-based examples, in tabular format, associated with corresponding graphic illustrations. Thus making the text vivid enough that the randomness properties gets reflected right from Columns of the Table to its graphic illustrations for both the RANDOM SCATTERS as well as their corresponding RANDOM MAGNITUDES , running parallel to each other. Here we could get the salient features of Part 2. Its details follow herein below.

As stated above, with initiation for obtaining RRNs in Chapter 4 it starts with the method of getting random numbers by approximation device in Chapter 5, the same follows to get the random numbers by filtration processes in Chapter 6. Each of the above two has to have four examples.

In Chapter 7 the author makes use of Bhaskaracharya's well known orthogonal trios 3,4 and 5 and extends his approach through several known Diophantine Equations. Thereby obtains RRNs by combining the same with arbitrarily chosen random numbers to have been resulted through Chapters 4,5 and 6.

Similarly for Chapter 8, author makes use of Hilbert space by choosing complex number necessarily having its root along with conjugate respectively for encrypting at source and for its decrypting at sink, (secretly kept at Argand Space) in combination with any large or small random number to get the targeted RRNs. Naturally each such numbers are independent of each other on account of Hilbertization to make Kapil's dictum a realizable reality.

The fifth and final approach to obtain RRNs from direct observations/ measurements of natures infinitely large creations have been demonstrated at Chapter 9.

The Chapter 10 being final chapter, the author presents systematic summary of the book as its EPILOUGE, that adds author's exhaustive studies and long experiences.

It is here that authors feel satisfied that RRNs could not only found but its forms and powers could be illustratively demonstrated.

But having such large scope for obtaining RRNs, brings forth new problem of keeping record of when and where any method and their amalgam have been used. Of course users have to maintain alphabetic index as where and when any method and its amalgam have been made use of. As we advance into Chapters we may find methodological to process the same.

10: Thereby follows Appendix containing four Appendices. Appendix 1 presents a picture of forest den just behind authors' house , where from he got inspired to draw random number directly by measuring features natural entities.

11: Who may read what?

Everyone who has acquired a copy of the book must read 'About the Book ' at back cover page and extension being continued elsewhere, preface and Chapter 1 in indicated sequence.

We have two classes of readers in mind. The first belongs to those one, not concerned with any background or reviews but directly on requirement of RRNs and their applications. The other being those interested knowing the philosophical paradox underling the use of PRNs.

The readers pertaining to class one may after reading up to Chapter 1 may like going to PART 2 but it would be better if one scans Chapter 4 also as it provides scope for obtaining RRNs from one's daily routine activities. That in PART 2 deals with the act of getting RRNs by any of the five processes contained in the text.

The other class of readers may like to read the entire text to have full understanding behind the reasons of choice of one over the other.

References:

1. Agnew, G.B.: (1988): Random Sources for Cryptographic Systems, Advances in Cryptology – EUROCRYPT '87 Proceedings , Springer –Verlag 1988 pp 77-81.
2. Griffith, R.T. H. and Max Muller Edt. (2007): RIGVEDA, The Oldest Divine Book, Vijay Goel, English-Hindi Publisher, Delhi, India
3. Hromkovic, J.: (2005): Design and Analysis of Randomized Algorithms (Introduction to Design Paradigms), Springer-Verlag, Berlin, Heidelberg, New York.
4. Maharshi, Kapil: (2007); SAKHYAKARIKA Edt. by Dr. R.S. Tripathy, Chowkhamba Press, Krishnadas Academy , Varanashi, UP,India
5. Vireswarananda, S. and Adidevananda, S.(1998): Badrayan's Brahma Sutras, Advait Ashrama, Calcutta, 700014.
6. Whyntie, T & Pugh,O: (2013) : Introducing Particle Physics, Icon Books, UK and the USA

Acknowledgments

At the age of past mid- nonagenarian one may think of writing a book is seldom possible without whole-hearted family co-operation. My two loving highly educated grand-daughters Monal (Smit Aakriti) and Sonal Aditi have always been helping hands, aiding my eyes, helping ears and whatnot and so has been their mother Smt. Susmita Ranjan for cares being taken for all that my elderly age needs to remain activated and continue working for the book. While on work for our memorable monograph I am away from other family members, that includes Sri Sanjeev Ghose (father of Rashi, my granddaughter-in-law) from whom I have been receiving encouraging words for the same I have been reciprocating my hearty thanks to all of them.

Also support received from Dr. Mukesh my youngest son, Div Rohit my eldest grandson and of Mohit yet another loving grandson at the stage of finalizing the manuscript and publication must get heartiest thanks. So, fully deserves Aparna (lovingly called Babli) for her health supporting dishes.

Apart from them, a philanthropic young friend Dr Govind Sharma, who has been EVER-READY a language collaborator for manuscript deserve appreciative thanks. Dr. Satish Sinha (IIT Dhanbad) as one of the top ranking executives and his family has been appreciative of all my academic ventures do deserve my blissful thanks.

Though last but not the least in importance for unabated support of the Shroff Publishers & Distributors, specially the Chairperson Sri. Aziz Shroff deserve my full-throated appreciative thanks for his instantaneous acceptance of all my manuscripts.

My blissful thanks get extended to Smt. Kiran Rastogi and Sri. Arif Shaikh to have come to our residence for error correction in our earlier publications. Now who remains to be thanked? Yes, the figure that emerges is that of Sri Sunil Shastri: perhaps without his unflinching attraction towards me or my writings, I would never been so AZIZ to Sri Aziz Shroff. To me, he is not yet retired but it is his phase of 're-treading of tires'. My best wishes to him and his family.

Professor N. C. DAS

Harshit Ranjan

About the Authors

Professor N. C. Das

Professor N. C. Das is former Professor-cum-Chief Scientist at the Department of Statistics and Computer Applications, Birsa Agricultural University, Ranchi, India. He has over seven decades of teaching- and research-experience in the field of Statistical Inference, Design of Experiments, Operations Research and Computer Science. Data-based modelling for prediction has been integral part of all above. These are now essential components of Data Science.

To his credit are World, Wide circulated undernoted publications:

1) Decision Processes by Using Bivariate Normal Quantile Pairs by Springer (2015). The WorldCat Identities till 2020 , the same was held nearly 300 WorldCat member libraries and considered Most Widely held works and with 3. 2k down loads. Further it has been as text of reference for 1) Decision making, 2) Multivariate analysis, 3) BUSINESS & ECONOMICS for Industrial Management, Management, Management Science and Organizational Behavior at Stanford University Libraries. US

2) **Data Science for Professionals:** Experimental Designs and Statistics with Least Resources, its two successive quick editions(2018 and 2022) as THE X-Team PUBLICATION by Shroff Publishers & Distributors, Navi Mumbai, along with FOREWORD by Professor A.C. ATKINSON of LONDON SCHOOL OF ECONOMICS.

   This publication drew attention of GCKE-2023 (The Global Congress of Knowledge Economy) held at Sapporo, Hokkaido JAPAN, to invite the Author to Chair/ Speaker for the event held in January 2023.

3) **Data Science for Prediction:** Simple Linear Innovative Model (SLIM) with Goodness of Fit for Testing Validity (GUD-FIT) Prediction (2021) by SPD, Navi-Mumbai. That is carving its own niche for its elegance, sleekness, simplicity and validity of prediction, minimizing the use LSE and discarding the application of R^2, traditionally being used. The same has been the source book to create demand for present one: the 'Simple Non-Linear Innovative Model for Prediction' with its pseudonym SINLIMP

4) **Data Science for Analytics:** Demystifying Essential of Statistics (2023) , is seen in great demand equally by empirical research workers and beginners of DATA SCIENCE learners..

5) For the Monographic Manual in your hand, get the requisite information at "About the Book" just here in attached separate.

All these serials from 2) to 5) have been published by SPD as noted at Sl. No. 2).

Thereafter his Mathematical Miscellanea consisting three undernoted papers are being submitted for quick publications in the best interest of the humanity. They are:

I) ARRIVAL OF LIFE corresponds to ADVENT OF DIGITS

II) A Perplexing Paradox:Stirling's Formula:Equivalent for n!, The Way out.

III) JOURNEY OF ZERO TO ITS POWER IT-SELF and so of ITS COMPLEMENTS

For his contribution to Data Science/ Statistics he has been honored by the Chancellor/ Governor of Jharkhand in 2022 and in 2023. Again for his publication "Entrepreneurial Designs for Start-Ups using Markandey's Digital Yantra (2023) he received ovation on annual day of the University from the Chancellor/ Hon'ble Governor of Jharkhand in June 2024. In December 2024, he was guest of honor at I. I. T. Mumbai where his sixth monograph named 'Simple Non-Linear Model for Prediction' was released.

He remained Academic Secretary-cum-Editor of the Bihar Journal of Mathematics during 1994-1998 and is currently as well as Founder Presidents of both, The Jharkhand Society of Mathematical Sciences and of Jharkhand Society of Statistics: Ranchi, which he has recently relinquished because of his age being past mid-aged nonagenarian.

Yet to the utter surprise he made incredible search to find the 'REAL RANDOM NUMBERS' (RRNs) hotly in search since last eight decades for optimal message encryption. For the same, till date, it has been virtually mandatory, to depend on Pseudo Random Numbers (PRNs) in spite of philosophical paradox existing against the same. Now with book in your hand such spate is expected to be over or at least to get minimized.

The best way to know an author is to know him from his PUBLICATIONS where he writes, refers other great writers and their ubiquitous publications as though they are still living and continue to remain so till their publications are in shelves of libraries of the World. That's the greatest award for an author.

He may be contacted on "profncdas@yahoo.com

*****

Harshit Ranjan

Harshit Ranjan is a final-year B.Tech Bioengineering student at Vellore Institute of Technology, Bhopal. With a strong foundation in computational tools and their applications in Bioengineering, He has made contributions to this book, focusing on the sections related to MATLAB graphics. His expertise lies in leveraging MATLAB for data Analysis, Simulation, and Visualization. His passion for blending technology with biological sciences drives his work, aiming to innovate in the field of Bioengineering and Computational Technology.
He is the youngest grandson of Professor N.C. Das, the first author of the book.

AUTHORS

TABLE OF CONTENTS

CHAPTER 2: Preliminary Review

CHAPTER 3: Investigating Review of the Creation and Usage of Pseudo RandomNumbers (PRNs)

CHAPTER 4: On Incredible Search for RRNs

PART 2: Obtaining Real Random Numbers (RRNs) From Approximations, Filterations and other Devices

CHAPTER 5

CHAPTER 6: On obtaining RRNs through Filtering

CHAPTER 10: Epilouge

PART 1

CHAPTER 1

Introduction

1.1: Random its requirement

It is useful in many ways, but only of its realistic forms obtainable mostly from natural world, incidence or instances. As it is there lies its vast warehouse but not of its imitations obtained from purported objectives as in practice. In fact it is there lies, the century old paradox.

1.2: Introductory Remarks

The English word 'random' did enter the arena of Statistics through a biological scientist Brown (1828) . The same then called 'Brownian Motion'. About which Whyntie and Pugh (2013) referred Einstein's (2005) paper that reported that the term 'random' appeared for the first time by Robert Brown (1828). He noticed that plant matter spat out of pollen grain suspended in water did 'randomly' jostle about under the lens of his microscope. The same could be seen to follow Gaussian probability distribution. There itself it got connotation of being haphazard without any reason or objective. Thus the concept of probability and random became jointly operative in probability and statistics.

1.3: Meaning of the word Random:

Since the book is about Random Numbers, it implicates to understand the meaning of word 'random'. The Dictionary meaning of this word is:

1): Want of definite aim or intention,

2): Something done, made or chosen without method or purpose,

3): - At random: without definite purpose or aim: haphazardly,

4): - adj.: Done or chosen without definite aim or for any deliberate purpose, and/or cause.

5): In statistics, it is to denote 'erratic'.

[Taken from The New International Webster's Comprehensive Dictionary of the English Language.]."

1.4: To face such meaning:

Facing such meaning, the author got utterly surprised when he could find that its similitude, as concept it existed much before. The same by Maharishi Kapil which has been excerpted by Pandey (1981) and Tripathy (1982) from Sankhyakarika 72,as herein under:

सप्त्यङ् किल ये अर्थास्तेऽर्थाः, कृतसंन्यास्य षष्टितन्त्राः।

आख्यायिका बृहद्हिता परवद्विवर्जिता तास्य चापि॥ ७२॥

Sapta-yang kile ye- arthas-te – arthah, krita-sanyasya shasti tantrah,

***Aakhya-yi-ka brihad-hitah paravad – vivarjita taasya-chapi** ||72||*

Import: The seventy practicable directives, bestowed through this scripture, pertain to Maharishi Kapil's shasti-tantra , meaning his 'Hexagonal System'. It is asserted therein, the same is required to be taken as, beyond any anticipation/apprehension impliedly beyond any intended objective, i.e. even perceived notion, at the same time making them beyond any exaggeration or suppression / attachment or detachment.

Indeed how great was Kapil who preached and aspired such virtue in human behaviour naturally, much before to-days parlance of the term 'random'.

Thus the second sentence of Kapil's couplet drives one to Maverick Effect which is just the synonym of Real Random, the subject matter of the text being written here.

Also such an epithet may very well be taken as 'true random' in present paradigm, sought after, since about a century. It is not contextual to explain 'Hexagonal System'.

However a short comment about Kapil's shasti-tantra (Hexagonal System) is being given at Appendix 3 as that relates to ternary GUNAS(sato-gun,rajo-gun and tamo-gun) every human being is necessarily required to pass through them.

1.5: The existing state:

It appears to have come in use to remove human bias in selection of some entities to form sample from a say a larger population. Yule and Kendall (1950) stated that human being is apt to be biased, hence the 'random' needs to be used in method of selection if it is required to fairly represent the features of targeted population.

Knuth (2000) refers Kendall and Babington Smith (1939) that with advent of computing machine, latter generated 1,00,000 random digits: a part from their Table has (only hundred) been annexed at Appendix 4 of this text as an example. This was followed by several such computer generated random number tables. In 1955, the RAND Corporation also published widely used of a million random digits with the help of another special device. Such spate continued till Marsaglia (1995).

Applied statisticians, other users of random number amongst several empirical scientists were using them without raising any objection on context of existing philosophical paradoxical ground.

The spate of objection became stronger with the application of such numbers in cryptographic use. Such generated random numbers were easily approachable to statisticians and empirical scientists.

In view of above dictionary meaning of the term random, it is awkward to note the sayings of Kendall (1947) and that of Feller (1950). While former wrote "In its colloquial use the word 'random' is applied to any method of choice which lacks aim or purpose. While he wrote its dictionary meaning that must not be regarded as mere 'colloquial' one.

Feller, on the other hand wrote,

'The word random is not well defined.'

Those statements are only to put forward plea for the use of the word 'random' in their own ways. Hence may not be taken as their conscientious expressions of such learned persons.

In order to wave out human bias as pointed out by Yule and Kendall (op cit) random sample selection was required to be resorted to. For such scheme of selection from larger population, the device of resorting for equal probability to all the entities became impliedly inescapable. The same could be executed as explained herein below:

Supposing from a population of size N, a sample of size n is required to be drawn. Therefore every entity of the population is required to be given chance i.e. probability = $1/({}^{N}C_{n})$.

Why so? It was so, for the reason that the step so taken could raise the status of Statistics (mere data representation) to empirical science i.e. the Science of Scientific Method. Indeed a lucrative opportunity. However the concept of giving equal probability was based on observational entities, be apprehended, to follow rectangular distribution. Be not so, the same in proportion to their density or frequency of occurrence. In fact, that was seldom possible when random numbers are required to be drawn from Natures' entities, the 'Nature' rarely follows rectangular distribution. A photograph of forest taken from the house of author's house has been placed at Appendix 1.

To facilitate the same several Random Number Tables were published as mentioned herein earlier.

1.6: Prospective Paradigm:

It is to be noted that wherever the term random appears as an adjective, like random event, random-variable, random-processes, randomized algorithms, randomised trials and random walk: the term there inescapably have been used purposively but not otherwise. However, we are apt to escape discussion about them, since for the present text our jurisdiction is limited to Random Number mainly.

Here we are concerned with its genesis, generation methods, philosophical paradox, including its rampant versus appropriate applications and usages.

As through such path, Data Science enters into Coding and/or Cryptography providing, safe encapsulation of information to be sent from well set source to targeted sink. Here the term cryptography is to mean the science of secret messages and their applications. The process is to transform plain text to cipher text at source and decrypt cipher text to plain text at sink.

1.7: How safe?,

Besides, who demands for what digital length?, how many are they ? ,and at what cost?, are the moot questions raised for the first time, that offers Clue for its fair/optimal appropriation as being raised, herein above.

Incidentally, they were Donald Knuth (2000), Press et al (1992), Nick Galbreath (2002), William Stallings (2005), Bruice Schneier & Niels Feguson (2003) and say Bruice Schneier (2010) to name a few, including thousands of references referred by each of them. They have advocated, enunciated and have proved some celebrated theorems for large Pseudo-Random Numbers as virtual substitute of True Random with little or almost negligible deliberations on moot questions raised herein above. One who attempts to write on random numbers must carefully read the writings of all above among which that of Knuth (op cit), who exhaustively has dealt the topic in his treatise. Quoting a reference of Hammersly & Handscombs (1964), they wrote it may not be necessarily required it to satisfy the statistical tests we have discussed. In fact inertia for such unethical avoidance or ignorance of such paradox, lied in generation of several Random Number Tables obtained earlier basically for statistical use primarily for sample selection.

1.8: Pseudo Random Numbers (PRNs)

Of course PRNs are important but for only few for their most important and most secret piece of information. Most of authors on this topics, have considered as though all information are equally important and they all need equal degree of encryption in spite of those are required to be complex and costly.

On the other hand, the market of 'information' in present scenario as large as this World including its Ocean, Mountains, Rivers and all its Living beings with all their problems and prospects, in times to come. The, rational thinking however is required to take care of them all for human races, own survival and growth.

1.9: The Operating Conflict:

The Conflict is between rampant use of costly and complex Pseudo-Random Numbers versus the genuine device, the Real Random Numbers. Those are required to be obtained, but must not be generated for any specifically defined purpose. That, due to philosophical paradox as already stated in its dictionary meaning. In fact such restriction must not be side tracked by merely changing terminology, 'Random' to 'Pseudo-Random'. Amongst these two, whether one may use former or latter the purpose is same.

1) Such changes must be regarded as 'ethical' one .The monograph in hand has to find a viable alternative, without needing the application of 'Pseudo-Random Number (PRN)'. As they are amply available through several voluminous published texts.
2) The solution has to come, only as simpler cheaper alternative to existing widely accepted texts based on PRN but no way in competition with them.
3) At its third stage, there had to be the selection of a treatment for being allocated to an experimental unit, say a plot. Again to nullify bias, random selection had to be resorted to.
4) One of the most important its aspects, is about testing medicinal effect by Hayes and Moulton's (2009) Cluster Randomised Trials, depending on genuine random numbers, say RRNs .
5) In fact the continuing use of PRNs is like performing of puppet dance in place of actual dance by an acclaimed dancer using RRNs, as also mentioned earlier.

1.10: Random Variable Defined:

Reza(1961) defines the term 'random variable' as a real- valued function defined over the sample of random experiment. Though restricting the same to keep real valued is quite natural as it is being used only in real format. Even though in various practical applications complex values of random variables are also considered. We shall be able to face such example in Chapter 8 of the text.

1.11: Existence of discrete similitude/congruence of Gauss's Normal Distribution.

As random number is usually discrete it naturally spans random variable a discrete probability function and/or distribution. Even in such format we can proceed for computation and inference by taking advantage of Yantrik discrete similitude/ congruence of Gauss's Normal Distribution. Refer Das (2023). The said distribution has number of advantages over former well known Normal Probability distribution. But the latter one is for continuous variables with limits extended from ($-\infty$ to $+\infty$) with its diminishing probabilities near large extreme values of x , the abscissa . Where as we now have its discrete similitude, which is symmetric , terminating finitely not needing central limit theorem. Its brief features are given at Appendix 2 of this text.

References

1. Das, N.C.:(2023): Entrepreneurial Designs for Start-ups, Shroff Publishers and Distributors Pvt. Ltd. Navi Mumbai 400705.

2. Feller. W.:(1950): Probability Theory and Applications, Vol. 1, Wiley India, New Delhi

3. Fisher, R. A. and Yates, F.:(1937): Statistical Tables, Oliver & Boyd , London

4. Galbreath, N.(2000): Cryptography for Database and Internet Applications, WILEY-dreamtech India Pvt. Ltd.

5. Hammersley, J. M. and Handscomb, D.C.:(1964): Monte Carlo Methods, Methuen, London

6. Hayes, R.J. and Moulton, L.H.(2009): Cluster Randomised Trials, C.R.C. Press, Boca Raton, London, New York

7. Kendall, M.G.(1943): The Advanced Theory of Statistics, Vol. 1, Charles Griffin & company Limited, London

8. Kendall, M.G. and Bobington Smith,B.(1939): Tables of Random Sampling Numbers, Cambridge Univ. Press, U.K.

9. Knuth, D.E.: (2000): The Art of Computer Programming, Vol.2: Semi-numerical Algorithms, 3rd. Edn.; Pearson Education Asia Pte. Ltd. , Indian Branch, Delhi 110092.

10. Pandey, O. P.:(1981): Sankhya-tatwa Kaomudi, Chaukhamba Saraswati Bhavan, Varanashi, U. P. India

11. Press, W. H., Teukolsky, S.A., Vellerling, W.T., and Flannery, B.P. (1992): Numerical Recipes in C, The Art of Scientific Computing, Second Edn. , Cambridge Univ. Press, U.K.

12. Reza, F.M.:(1961): An Introduction to Information Theory, McGraw Hill , New York, Toronto, London , Tokyo

13. Schneier, B.(2010): Applied Cryptography, 2 nd.Edn., Wiley India Pvt. Ltd. New Delhi.

14. Schneier, B. and Fergusan, N. : (2003): Practical Cryptography, Wiley – dreamtech India Ltd. New Delhi.

15. Stallings, W. (2005): Cryptography and Network Security Principles and Practices, Pearson Education, Delhi

16. Tippet , L. H.C. (1927) : On 40,000 Random Digits.

17. Tripathi, R.:(1982): Sankhyakarika. Chowkhamba Krishnadas Academy , Varanashi. U.P. India

18. Turing, A.M.:(1955): The RAND Corporation published million Random Digits.

19. Typhoon International Corporation,(2004): The New International Webster's Dictionary of the English Language, 801 12th . Avenue, South Suite 400, Naples, Florida 341 02.

20. Yule, G.U. and Kendall, M.G.:(1950): An Introduction to the Theory of Statistics, 4th. Edn., Charles Griffin, London U. K.;

CHAPTER 2

Preliminary Review

2.1: Pseudo-Random Numbers (PRNs):

In Chapter 1 itself we have outlined about paradoxical problem that are required to be faced for raising alternative to 'Pseudo-Random Numbers' (PRN). Almost all its proposers and developers were fully aware of the philosophical paradox but in spite of those, they hunched towards the same, ever since its indispensable requirement has been felt for its cryptographic and other applications.

2.2: On Knuth's and those of Press et al, s Contributions:

As referred in Chapter 1, Knuth (2000) on making one of the longest review, stooped to such usage of PRN on account of its popularity and sizable international market it had already occupied till then. Also, Press et al (1997) in their treatise while deliberating on random wrote:

"It may seem perverse to use computer, that most precise and deterministic of all machines conceived by human mind to produce 'random' numbers. More than perverse, it may seem to be a conceptual impossibility. Any program, after all, will produce output that is entirely predictable, hence not truly random."

To repeat even after having such view about PRN, but for its inescapable need, quoting Knuth (op cit) they also stooped to the usage of PRN. As its viable alternative could not be apprehended then such spate had to be tolerated and hyped.

2.3: On Stallings Contribution:

The well known protagonists of current principles and methodologies of Cryptography, Stallings (2005), having deep feeling for such philosophical paradox, impelled to express inclination towards possible methods to procure RRNs. Thereby he guessed the same to be probably obtainable from physical noise, pulse detectors of ionizing radiation events, gas discharge tubes and leaky capacitors are some potential sources but finds them not implementable in larger scale use.

2.4: On opinion expressed by Schneier:

Similarly Schneier(2010) writes and quoted Agnew's proposition:

" The best way to collect a large number of random bits is to tap the natural randomness of the real world and adds by quoting Agnew's(1988) propositions as below:

'It is metal insulator semi-conduction capacitors, two of them kept in close proximity and the random bit is a function the difference in charge of the two.' Besides he cites some more examples from similar situations.

At the same time they felt that their such propositions, are rare and far between impliedly for commercial purpose.

From above examples, it appears that concerned scientists were and even now are conscious of such philosophical paradox and for its viable alternatives, search is on.

The challenge therefore is of its (such methods) having been got created or generated, independent of any intended usage. At the same time not needing so much mathematically oriented and/or statistically tested, random numbers.

2.5: The Scope for Approximation Approach:

It is from Stallings(op cit) and Schneier(op cit), the author could get the clue for the purposeful use of his large scale iterative computational experiments as approximants to reach any prefixed scalar target: as its differences or residues (called error), are neither related to, nor obligatory for the same to fulfil any other mundane objective. It must be made clear as to what it means, is the term 'error' in computational perspective. To be clearer, let the same be obtained by Approximation Algorithm being noted here:

1) Choose any numerical real target to be approximated,
2) Select any suitable iterative technique (say repeated halving) to reach the set target value, that within prescribed limit of small (e^), say epsilon number.
3) Continue the process till it reaches the target within the prescribed limit.
4) As it reaches the target within (+/- e^), get the output, the tenable approximate.
5) Abs (target - approximate) and/ or (target - approximate) with their sign as it comes, is to be taken as 'Real Random Numbers', the RRNs.

But present paradigm shift it is being largely used/ understood, as discrepancies to meet any sought after numerical target. Thus it has to have some digital/numerical magnitude; however small it may require being so.

Author felt then that such processes anticipate results in 'fairly long series of random walk'. Their magnitudes presented, in tabular format for their large scale use. Those be random, no doubt but not to be subjected to any statistical tests. As they are not the least directed by any mathematical and/or any other object related principles. They are just errors, but restricted to remain within two limits: so that they do not fall victim of large deviations. They are to be large and bi-directional, but to remain only between finite- negative to finite positive, so far. How that is to be achieved, may be found in the text. But it is liable to change its sign and magnitude without any impact on its being random. Though sometimes positive and negative of equal magnitude are seen to acquire identical values but such repetitions have little impact on their random nature.

2.6: Sayings of Einstein:

I may like to close here by quoting the sayings of Einstein (2005):

"That the Lord should play the dice, all right; but that He should gamble according to definite rules that, is beyond Me.": Quoted from CSI Communications, Mumbai, India December 2005, Front Cover Page.

2.7: Authors' Approach:

That is why we have obtained such number to come of its own without any higher mathematics and the same time not needing any statistical test. Behold the Nature, it as diverse as congruent/ similitude, sometimes though rarely so even identical amongst the said diversity but no one may say, 'THEY ARE NOT RANDOM and/ or their' MEASURABLE FEATURES ARE NOT RANDOM VARIABLS'. To authenticate our statement, we are presenting at Appendix 1, some photographs taken by associate author from our house-roof, as has been stated in Chapter 1 itself.

2.8: Chapter conclusion:

To conclude it is argued that if measurable features drawn from nature are random variable, we may have some methods to convert or draw desired random number out of such random variable. Though there exist its possibility but reversion may or may not be possible. For example though curd is milk product, it not possible to revert the same to milk. Also even if reverted it cannot be equally distributed. However for coding and/or encryption such features are not required. Only its largeness may be necessary for highly confidential communication.

References

1. Agnew, G.B.:(1988): Random Sources for Cryptographic Systems, Advances in Cryptology, EUROCRYPT-87, Proceedings, Springer-Verlag, Germany
2. Aitken, A.C.:(1949): Statistical Mathematics, Sixth Edn., Oliver and Boyd Ltd., Edinburg and London, UK
3. Einstein, A.(2005): CSI , Computer Society of India December 2005, Mumbai, India
4. Elderton, Sir, W.P. , (1938): Frequency Curves and Correlation, Cambridge Univ. Press. U.K.
5. Kendall, M.G.:(1947): The Advanced Theory of Statistics, Vol.1, Third Edn., Charles Griffin & Company Ltd., London, UK
6. Knuth, D.E.: (2000): The Art of Computer Programming: Semi-numerical Algorithms, Vol. 2, 3rd. Edn. Published by Addison Wesley Longman (Singapore) Pte. Ltd.-Indian Branch, Delhi 110092
7. Pearson, K.(1895): Skew variation in homogeneous materials; Philosophical Trans. A 186: 343
8. Press, W.H. ,Teukolsky, S.A., Vellering, W.T. and Flannery,B.P. (1992): Numerical Recipes in C, The Art of Scientific Computing, Second Edn. , Cambridge Univ. Press, U.K.
9. Scheiener, B. (2010): Applied Cryptography, 2nd. Edn., Wiley India Pvt. Ltd. New Delhi. India
10. Stallings, W. T. (2005): Cryptography and Network Security Principles and Practices: Pearson Education, Delhi

CHAPTER 3

Investigating Review of the Creation and the Usage of Pseudo-Random Numbers (PRNs)

3.1: Creation of PRNs:

Pseudo Random Number Generation (PRNG) was first proposed by Lehmer (1949 & 1951) in his classic symposium paper published by Harvard University Press. Where from its conceptualization with Modular Arithmetic began. The prime need for encryption was realization of large integer number those being non-divisible and its residues being necessarily obtainable, as reported by Stallings (2005) even quoted by Hamming (1991).

The Chapter is primarily meant to discuss Pseudo-Random Number which is widely in use and accepted as viable substitute of 'true random numbers', as the latter believably could not have been generated. Authors anticipate, it is possible to generate the same.

The method to create 'Pseudo Random Number' has been primarily based on two principles.

Those are

1) Use of large prime numbers and
2) Modular arithmetic, both as complement to each other.

For coding and/or encryption the basic requirement is of large number, as larger the number, harder it is, more time consuming or more costly the same, to break up the cipher text, meant to eliminate or minimize, unauthorized decoding or de-encrypting the text message.

The application of modular arithmetic results in yielding non-zero residue which is needed for Pseudo Random Number generation. It is the prime number that necessarily yields non-zero residues. That also are amongst unavoidable requirements. Hence to satisfy the twin requirements, large prime number and modular arithmetic, turn out to be indispensable choice.

3.2: Early Approach for creation of PRNs:

Much before such recent attempt, Tippet (1927) and Kendall & Babington-smith (1939) for their respective sizes 40,000 and 100,000 of so called random digits got generated. They were all object oriented and machine produced, hence not worthy to be called 'True or Real Random Numbers'. To exemplify, it is apt to attach one such table, an initial part of Kendall & Babington Smith's (op cit) Table, is being attached at Appendix 4 with only hundred numbers only.

In face of well raised philosophical paradox, any alternative in use, in violation of the same needed to have been damped as unethical. Hence search for RRNs should have been perused. Instead of such effort, machine or computer generated referred in preceding paragraph, continued unabated.

Facing such situation and increasing market demand, number theorists were inclined to develop algorithms and computer programs for creating faster and larger pseudo random numbers. The same appears to have initiated by Lehmar (1949) as quoted by Knuth(1998). There after by Press et al (1997) who apart from algorithm, they presented their coded program in C- language. They have expressed objection on its use on the basis of existing philosophical paradox, but stoops to accept the same, as it was widely accepted and remained in practice for drawing samples from larger population.

3.3: Writings about 'random' by some renowned authors:

1) Knuth (1998 & 2000) writes,

"People who think about this topic invariably get into philosophical discussion about the word 'random' means. In a sense, there is no such thing as random number; for example is 2 a random number? Rather, we speak a sequence of independent random numbers with specified distribution and this means loosely that each number was obtained merely by chance, having nothing to do with other numbers of the sequence, and that each number has a specified probability of falling in a given range of values. A uniform distribution on a finite set of numbers is one, in which each possible number is equally probable. A distribution is generally understood to be uniform unless some other distribution is specifically mentioned,"

2) Dagpunar (1988) writes:

Following Knuth(op cit), Dagpunar (op cit) wrote a text on "Principles for generation of Random Variable". Starting with, random number sequence, discussed also empirical tests for the same. Thereby gave FORTRAN 77 routines for generating variables from selected distributions: those primarily being Normal, Gamma, Beta and Chi-Squired, 't' and F-distributions. Apart from above he considered for the purpose Poisson, Binomial and Negative binomial distributions. That way it could be a very useful text. But incidentally his approaches have been through Pseudo Random Variable Generation (PRNG).

3) Galbreath (2002) and others:

Defines the same as, it is required to be generated from random source, which generates numbers that are independent of all previous numbers before it. That there exists every subsequence of numbers and also that any subsequence of say length n occurs equally as often as others sequence of length n. That implies giving equal chance (i.e. probability) to all possible to all subsequence of size n. Here the ticklish term is very choice or its initial choice of 'random source'.

Thereafter he switched to generation of Pseudo Random Numbers (PRNG), and various tests to prove its worthiness of being random.

Following him soon Schneier and Fergusion (2003): They relatively giving long discourse on immense difficulties in getting real random numbers and hence pleaded for adoption for PRNG.

3.4: Contribution by Number Theorists:

Almost simultaneously several number theorist got raised to contribute for such ventures but almost all of them for Pseudo random numbers. Some such publication as available to author, were those by Herkommer(1999), Hankerson et al (2001), Koblitz(1997) and Lewand(2000). All the four publications are masterpieces of their contents and styles. But the author here has preferred more the one pertaining to Hankerson et al's, not only because it has seven co-authors, but also it is slick and simple and covers both aspects Coding Theory and Cryptography .

However that yielded scenario, where generation of real random number and consideration for associated philosophical paradox, pretended to be ignored. Some of them did consider bringing concept of probability to their contributions. Also none of these four attached much importance to the term 'random': much less about its generations.

3.5: Probability and Random vis-a-vis Pseudo Random:

Earlier we referred Feller (1972) for the term 'random', who wrote "The word 'random' is not well defined, but when applied to samples for selections it has unique meaning. The term random choice is meant to imply that all outcomes implying selection of entities of sample drawn from hypothetically larger population are equally probable."

Loeve (1963) in his treatise on 'Probability Theory' profusely made use of the term and every event considered in probability theory, is to be treated as random event.

However neither of two did talk about the term 'Pseudo Random' much less of its generation and its application any- where for probability world.

Winston(2004) in well written text on Probability Models also adopted and advocated the use of modular arithmetic for generating random number as being done by others in method of Monte Carlo sampling for simulations. He also gave an example on Agricultural Planning under uncertainty, considering weather being highly uncertain. Hence using such random number, where farmer could maximize his expected profit.

Hromkovic (2005) virtually made no reference for the term 'Pseudo Random'.

Ludeman (2010) in his text on 'Random Processes' made no reference about any specific method for random selection, like the term Pseudo-Random Numbers or any other method for same. Thus every one of them treated the term 'random' as 'random' there being no necessity of clarifying, while for choice amongst them, which one to use.

3.6: Latter attempts:

Referring to standard text on 'Cryptography and Network Security by Stallings (2005), he at first writes on sources of Random Numbers. Where clearly writes, 'Sources of true random numbers are hard to come by'. However he had some realistic hunches about its existence. Wherein he indicates 'Physical noise generation such as pulse detectors of ionising radiation events, gas discharge tubes, and leaky capacitors are one potential source. He however soon deviates from the same on ground that such devices of limited utility in network applications and lends towards Pseudo random numbers generated by existing sources.

The other well known text on 'Applied Cryptography by Schneier (2010), over writing his earlier publication writes 'A random –sequence generator's cannot be produced. Further he writes 'There is a large philosophical debate about whether any of these techniques actually produces real random bits. But soon he refers to Agnew (1988) for his hunches as herein excerpted below:

'It is a metal insulator semi-conduction capacitors. The two of them are placed in close proximity, and the random bit is a function of difference between the two. Another random number generator generates a random-bit stream based on the frequency instability in a free-running oscillator. Agnew (opcit) gave some more examples real life natural instances. Thereafter Schneier (2010) in his paragraph on 'Distilling Randomness' writes 'In general, the best way to generate random numbers is to find a whole lot of seemingly random events and distil randomness from them. This randomness can then be stored in a pool of reservoir that can be drawn as when needed.

The above written paragraph becomes the eye-opener for providing clue to the author in his search for Real Random Numbers (RRNs).

Yes, there are some repetitions to what have been talked about in Chapter 2 but that is quite natural as here we deal the same with details about the same as on Investigation in depth.

3.7: Relevance, for creation of RRNs:

What matters is its (RRNs) creation for purposive usage. In fact philosophical paradox, restricts to its unethical usage that any 'random entity must not relate to its purposive usage'. In fact it is that intent, the author is harbouring at the same time, inspiring him for search of RRNs.

3.8: On its usage for Medicinal Purpose

One of the most important aspect of using concept of randomization dependent on random number lies in its use for Cluster Randomised Trials for testing medicinal effect for mass scale (for both preventive and curative) uses. There it-self, its non-genuine (artificially generated) form should have been strictly prohibited.

However perchance if the same gets created for non-purposive usages as any by- product, meant to have been created for unconcerned, unrelated product or theme, in terms of magnitude or measure, it must by implication qualify for being purely or truly random number. Such a random number may not be required to satisfy for any condition or artificially created test or statistically formulated test. On the other hand, the same automatically generate the theme, 'random is random', nothing else.

3.9: Power of Random, an Example:

We have an Anecdote to show the power of the random for reduction in weighing of balls which is being excerpted from Das(2019 and 2022).

Anecdote: Suppose there exist, eight balls of same size, shape and of colour out of which seven of those have same weight also: but one them is heavier. Also there is available a traditional balance (having fulcrum, beam and two weighing pans tagged to beam but no weight measuring units). It is required to use balance only twice and detect the heavier one. Usually it would take three weightings. But using random selection and consequential inference, as explained here only two could do.

1) Take six balls at random out of eight: divide three balls for each of the two pans. If both sides are equal, then put all of them aside, as heavier one is not in any of them. The second weighing of remaining two balls, could easily detect the target ball.
2) In case the first weighing is un-equal, then select such three which are heavy. Amongst such three select two randomly: if they are equal, the remaining one is heavy one.
3) In case they be unequal then heavy one is the target ball.

Thus using randomised selection twice, the two weightings could solve the problem without any weight measuring units. The power of random could very easily be demonstrated.

3.10: On Theorem of Motwani and Ragavan(2004):

I am the senior author who has to submit that: "The Chapter 12 of Computer Science Handbook has been surfed by me, which is due to Motwani and Raghavan (2004) which also refers to their esteemed publication (1995) and several other related publications. Its most impressive section is Section 12.4 where they through Theorem 12.3 have obtained *'Given any instance of* T_k *, the expected number of steps for preceding randomized is at the most* 3^k*.' : thereby argued 'Because* $n= 4^k$ *, the expected running time of our randomized algorithm is* $n^{\log_4 3}$, which we bound by $n^{0.973}$. Thus the expected number of steps is smaller than the worst case for any deterministic algorithm. Further they comment,' Note that this is a Las Vegas algorithm and always produces the correct answer.' "

We express our regard towards their finding with due honour and respect for their celebrated theorem. Incidentally however the same is within the ambit of philosophical paradox, also fall short of the concept of Maharshi Kapil about his envisioning aspiration of 'SIMLITUDE/ CONGRUENCE' to modern term 'RANDOM' expressed several millennium years before.

3.11: Random and Algorithms:

Virtually on completion of the book, it came to my memory that random have had some role in Algorithms also. We should refer them also. We have only three books in our library. They were of Coreman et al (2009), Johnsonbaugh & Schaefer (2009) and of Goodrich et al (2013). All the three sets of authors considered its equally probable aspects to have been generated by computer generated methods (PRNs). They did not refer RRNs. However they dealt on randomized quick-sort algorithms.

3.12: Chapter Conclusion:

The Chapter eventually closes with conclusion that PRNG may continue to be resorted for large scale use in Cryptography, if any of the proposed devices or method fails to satisfy user's or market demand but in full realization of the fact that the same false short of philosophical paradox, hence unethical. It is like raising a tower on week foundation.

References

1. Agnew, G.B.:(1998): Random Sources of Cryptographic System, Advances in Cryptology- EUROCRYPT'87, Proceedings : Springer Verlag 1988, Pp 77-81.

2. Coremen, T.H., Leiserson, C. E. and Rivest. R. L.:(1990) , Introduction to Algorithms, Prentice Hall of India, New Delhi

3. Dagpunar, J. : (1988): Principles of Random Variate Generation, Clarendon Press , Oxford, U.K.

4. Das, N.C. :(2019, 2022): Data Science for Professionals (Experimental Designs and Statistics with Least Resources), Second Eds. , Shroff Publishers & Distributors Pvt. Ltd. Navi Mumbai,400705.

5. Feller, W./:(1972): An Introduction to Probability Theory and its Applications, Wiley Eastern Pvt., Ltd. New Delhi

6. Galbreath, N.: (2002): Cryptography for Database and Internet Applications, WILEY-dreamtech India Pvt. Ltd. New Delhi

7. Goodrich, M. T., Tamassia, R and Goldwasser, M. H.: (2013): Data Structures & Algorithms, John Wiley & Sons, New York, US

8. Hankerson, D. R.; Hoffman, D.G.; Leonard, D.A.; Linder, C.C.; Phelps, K.T.; Rodger, C.A. and Wall, J.R.:(2001): Coding and the Essentials; Second Edn., Marcel Dekker, Inc., New York, US

9. Hayes, R.J. and Moulton, L.H.(2009): Cluster Randomised Trials, CRC Press , Boca Raton, London, New York

10. Herkommer, M.A.: (1999): Number Theory: A Programmer's Guide, Mcgraw-Hill, New York, New Delhi

11. Hromkovic, J. :(2005): Design and Analysis of Randomized Algorithms: (Introduction to Design Paradigms), Springer, Berlin, Heidelberg, New York. US

12. Johnsonbaugh, R. and Schaefer, M.(2009): Algorithms, Pearson Education, Delhi, India

13. Kendall, M.G. and Bobington Smith, B.(1939): Tables of Random Sampling Numbers, Cambridge Univ. Press, UK

14. Knuth, D.(2000): The Art of Computer Programming, Vol. 2; Semi-numerical Algorithms, Koblitz, N.:(1997): A Course in Number Theory, Second Edn. , Spinger , Berlin, Germany

15. Lehmer, D.(1949& 1951):' Mathematical Methods of Large-Scale Computing' Proceedings of 2nd. Symposium on Large –Scale Digital Calculating Machinery: Cambridge: Harvard University Press.

16. Lewand, R.E.(2000): Cryptological Mathematics, The Mathematics Association America, MAA Science Centre, Washington D.C, U.S

17. Loeve, M.:(1962): Probability Theory , Von Nostrand, New York, US

18. Ludeman, L.C.:(2010): Random Processes, Wiley India Pvt. Ltd. New Delhi

19. Motwani, R. and Raghavan, P. (2004): Randomized Algorithms, Chapt. 12 of Computer Science Handbook. CRC Press, LLC

20. Motwani, R. and Raghvan, P.(1995): Randomized Algorithms, Cambridge Univ. Press, New York.

21. Schneier, B. :(2010): Applied Cryptography , Second Edn., Wiley India, Delhi.

22. Schneier, B. and Fergusan, N.:(2003): Practical Cryptography, Wiley-deamtech India Pvt. Ltd. , Delhi

23. Stallings, W. :(2005): Cryptography and Network Security Principles and Practices, Third Edn. ,Pearson Education, Delhi, India

24. Tippett, L. H. C.:(1927): Random Sampling Numbers (Tracts for Computers No. 24), Cambridge Univ. Press. UK.

25. Winston, W.S.:(2004): Introduction to Probability Models, Operations Research Vol.2, 4th. Edn., Thomson Course Technology, Eswar Press, Australia, U.K. and U.S.

CHAPTER 4

Incredible Search for RRNs

4.1: Introductory note:

While referring to Schneier(2010) at Section 3.6 of the text Chapter, the Author got the hunches that it is possible to obtain and generate Real Random Numbers (RRNs) from natural events or instances: obviously from theirs features, counted or measured in digits and/ or numbers. Also that such quantitative measures should not be subjective or any objective pursuit, what so ever. Clearly it may arise accidentally or eventually in counts or measures of some other pursuit, as their by-products.

That inspired the author, to make search for the same.

4.2: The Search begins:

In fact about search, we have already mentioned at Section 1.4 of Chapter 1 even before surfing Schneier (op cit) the Author have had the hunches that (RRNs) could be found from features of the Nature, but not necessarily from Modular Arithmetic combined with Large Prime Numbers as being perused presently.

4.3: How and from where from to fetch RRNs?

In referred texts above we had have its (true random numbers) genesis, but not how to get them. Whenever any concept emerges, then it must have found use somewhere. There is ancient maxim 'if there exist some effect, then there must be cause behind rather before the same'. It is near impossibility to create something without its need or requirement somewhere.

Badrayan, the philosophical name of Vedabyas in his ***Brahmsutra – Vedaanta-Darsan wrote***

भावे चोपलब्धे: and ना भावअनुपलब्धे:

Bhave-cha-uplabdhe and Na-bhavo unupabdhe

Import: If effect can be realized, then there must be cause before. Also if the effect cannot be realized then the cause cannot be detected.

It has already been quoted earlier also.

4.4: Authors' Hype:

The Author had hype 'whatever mankind is required to find something which is not of their objective-usages, but merely for throwing away or may not be of any purposeful usage, or say as its by-product only. In process of thinking he found many such instances but generally not taken care of. To find something very commonly being done, he found husk of the grains.

A medieval philosopher-poet Rahiman's concept of error (thotha) derivable from a farm lady's action:

सार सार को गहि रहै, थोथा देई उड़ाय।

रहिमन सार को गहि रहो, थोथा देइ उड़ाय॥

Saar saar ko gahi diyo , Rahiman Thotha deyi uraai:
Rahiman saar ko gahi- raho, Thotha deyi urrai .

Import: A farm-lady was found by the poet Rahiman that she gets crushed first grain-kernels, naturally canopied with husks. Then she winnows grains and blows-out the husks with wind, to stack the grain thus separated of husks. Her prime objective was to get grains, purified of husks. What happens to husks and what is its quantity meaning is weight, volume, its energy value in Kg/ Calorie, were not her objectives. Here husks, its attributes therefore could be treated as 'error'. Those may convenient1y and abundantly gotten as 'real random numbers' on retaining digits or numbers only, bereft with its, of what they are or were . They are naturally bound to be erratic or say 'haphazard'. The term Kendall (1947) used to mean later on as 'random'.

Also Reza(1961) in Section 12-2 of his classic text on Information Theory while writing on 'error' states "The error E associated with a decision scheme can conveniently be considered as a 'random' variable ....". Also thereafter he excerpts Feinstein(1959) theorem for the its 'Equivocation Probability' based on convexity property of the entropy function'. It would be interesting to find such property of error well demonstrated in numerical examples of Chapter 5 of this text with their graphic representations.

4.5: Its (RRNs) Clue:

It is from here we have the clue as to wherefrom such Real Random Numbers (RRNs) are naturally obtainable. The measurements and magnitudes of terms or words that are commonly used for things or objects like: *remains, remnants, remainder, residue, rubbish and rudiments* are likely to be sources of RRNs. Some terms like noise or difference are also used instead of terms shown herein italics. Their uses vary from discipline to discipline and/or context to context. What else are wanted is to know their magnitudes in terms of digits or numbers. Those may be real or even complex. They may be thus very large or small as one needs for its use. Not necessarily large prime numbers as mandatory requirement for PRNs .

4.6: Factual Statements about it (RRN):

It is fact that it may not be possible to take measurements even observations of many such events or instances, meaning all above shown in italics, but there exits vast scope for wherever it is possible.

1) Suppose you get lists for provisions procured from market with items name, its quantities and prices all in digits or numbers, forsake their scale and retain only its counts or measures in digits or numbers, take them to be stacked as RRNs .
2) Suppose you have received telephone calls from say fifteen persons, retain only any say two digits of each such numbers for your RRNs to be used tomorrow.
3) Suppose you draw some amount from bank and disburse the same to three of them, retain only one or two digits from each such amounts of them to be used after say three days after from now including from amount you have drawn.
4) Now your eyes, ears and your mind are opened, you may easily choose as to which of your remains, residues,...., and rudiments may easily be counted, observed, measured or weighed and retained which may include household to kitchen waste, the same as your RRNs today.
5) In case you feel your inability for above suggestions, still you have many options. For example plants or creepers in your house, you may arbitrarily reach some its twigs and measure their leaf areas or even weights of those leafs that have shed off plants on maturity, keeping in record only their digits or numbers not the scales.

4.7: Advantages of its features:

1) For such measured properties, as randomness no test is required.
2) Such random is purely 'real random', its feature cannot be got changed, even if it gets multiplied by any scalar.

3) Even sums and/ or differences of say two of them are bound to remain random.

4) Every communicative-congenial group may have their own sets of such random numbers.

5) Generation of such random numbers are neither demanding for larger prime numbers nor for modular-arithmetic.

6) Scope for getting 'real random numbers' (RRNs)

7) Apart from above naturally obtainable real random numbers (RRNs), we have huge sources of getting the RRNs from sources of, 'Approximations' pertaining to Numerical -Mathematics or Statistics, where one has to have a scalar target for which approximation (called approximants) is needed to its desired accuracy: say given magnitude of error epsilon é . Having reached or exceeded the same, user gets satisfied that satisfactory approximation to the fixed scalar quantity has been reached. Differences between such scalar, to its approximants obtained at each iteration, also provide RRNs. However, those are of generally of smaller magnitude which may be enlarged, as desired by multiplying the same by large number. In fact the author has good stock of the same. Some of them shall be used to generate sample Tables of RRNs. But they must not be used for anyone for the purpose of encryption, as obviously they stand exposed to public , implying to have been published.

4.8: Direction towards RRNs:

The above thoughts direct, the author to search for truly random numbers. It is just like searching a baby who is already in mother's lap.

In case observations, measurements or realizations are taken to be random variables, then such observation is say x, therefore $u = x - v$ is the filter and v is known as 'noise' , the $v = x-u$ is RRN quantity.

1) Schneier (2010) has referred 1653 publications on generation of Pseudo Random Numbers (PRNs). He makes statement that 'The best way to collect a large number random bits is to tap the natural randomness of the real world.

2) Stallings (2005) also made statement that RRNs are obtainable from physical noise generation such as pulse detectors of ionising radiation events, gas discharge tubes and leaky capacitors. However, he assigns it to be the potential sources that have to have limited utility.

3) Such RRNs may or may not be always large, rather is length may depend on importance of its confidentiality for encryption. That may optimize the cost on encryption.

It is such statements which initiated thought that it is possible get RRNs from various natural sources.

4) Some photographs of natural scenes as at Appendix 1 may bring to light that it is Nature which conceal as well reveals its features to be the Real Random Numbers (RRNs).

4.9: Caveat and Caution for its (RRNs) uses:

The above theme draws towards attention that magnitude of anything like 'error' or its any variants may be RRNs. But that is very subjective in the sense that what is classed as 'error' may be a 'reality' if context changes and vice-versa.

1) For example if you are to consume (as eatable) rice, its husk-less form is 'truth' and husk is to be considered as 'error'. Otherwise 'husk' is required to used for getting 'oil' or 'paper' , then it is the 'truth' and edible husk-less is the 'error'. Thus one has to be careful as to what is what and when or then. Also in form of paddy it is reality or say truth when it to sown for growing plant and its earlier forms are to be considered as 'errors' or say, 'untruth'.

2) Moot question raised here is that 'there is huge refuses as war destructives or bull- dozer demolitions: can its measurements if taken as RRNs?

Its answer is No: Why so? As, they are being, objectively sought after though with destructive motive or panel action.

4.10: On expansion of its(RRNs) Scope:

Keeping in view there exists immensely large scope to obtain RRN of any size for various uses by individual and/or groups and societies from their own resources to be kept completely secrete to exchangeable between source to sink. Also they may throw away better burn only after single use.

4.11: On variations of its (RRN) requirements:

1) The requirement of RRNs may be vary for varying occasions. The one that you need is very near your problematic requirement in shape of error close to such problems.

2) No individual number may be considered as random. It has to come in sequence with other such numbers. Such sequence, though a discrete-assemblage be such that they be all independent as none of them gives or takes anything or thought of one another like favour or disfavour.

3) Further such a sequence cannot be less than three: larger than that it may show better scatter, haphazard look or say randomness gets reflected. Usually it should be larger than 10 but not more than thirty.

4) Much more can be had on association and/ dependence of one over the other, 'error' and 'random' from extensive contributions on them from voluminous texts by Jelinek(1969) and Huffman and Pless(2007) as both relates to Information Theory and Error-Correcting Codes and similar other contributions not within the reach of authors of this texts.

5) For generality and operations we assume that all the basic assumptions required in algebra may be applicable here also, yet the random number so obtained will continue to remain random number except that two identical random numbers when subtracted from each other would be zero, or divided with one to other, one gets '1' which may not be regarded as random number. Similar other operations may not yield viable number.

6) As regards search for random numbers, it is to remain limited to discrete number as most of the events and/or entities that we encounter happen to be discrete and finite, whereas both continuity and infinity are merely phenomenon to have been supposed mainly for mathematical convenience.

4.12: Foiling any Adversary:

With lot of Choices for obtaining RRNs, no doubt any adversaries attempt could be foiled. The same to keep confidentially as to which RRNs has been used where and when may itself would be reasonably difficult. For such problems one may refer Motwani and Raghavan (2004) and (1995), to have Randomized Data Structures to store and quick access to those. Better to have one's own creations using lexicographic style separately for business group, family group and any other groups to be kept in complete secrecy or limited so. If not done there is danger of falling into own created cobweb. It is therefore better to follow some processes being created, explained, demonstrated and illustrated in Chapters of Part 2.

4.13 About Next Chapter:

With next Chapter 5, Part 2 on Application gets started. There from we obtain RRNs through different approaches giving examples with their graphic illustrations. The RRNs obtained therein shall not be of cryptographic usage as they remain exposed through publication of this text. Hence their confidentiality gets lost. However Authors have good stock of fresh RRNs. They may be made available on request to authors.

References

1. Badrayan (1998) : Brahma-sutra, Advaita Ashram, 5 Dehi Entally Road, Kolkata -700014, India

2. Huffman , W.C.:(2007): Fundamentals of Error-Correcting Codes, Cambridge Univ. Press, London and New York

3. Jelinek, F.(1969): Probabilistic Information Theory: Discrete and Memoryless Models, McGraw-Hill Book Company, New York

4. Kendall, M.G.:(1947): The Advanced Theory of Statistics, Vol.1, Third Edn., Charles Griffin & Company Ltd. London.

5. Motwani, R. and Raghavan, P:(1995): Randomized Algorithms , Camb. Univ. Press, New York.

6. ---------------------------------------. (2004):Randomized Algorithms. Chapter 12of Computer Science Handbook, CRC. Press, Ltd.

7. Reza, F. M.:(1961): An Introduction to Information Theory, McGraw Hill, New York

8. Schneier, B.:(2010): Applied Cryptography, Second Edn., Wiley India, New Delhi

9. Stallings, W.(2005):Cryptography and Network Security, Third Edn., Pearson Education, Delhi- 110092.

PART 2

Obtaining Real Random Numbers (RRNs) From Approximations, Filterations and other Devices

CHAPTER 5

RRN has to remain INDEPNDENT & UNCONCERNED

with any of the METHODS used for PRNs.

Yes, there exists RRN and its SCATTER , we are to gain its MAGNITUDE.

5.1: Obtaining Real Random Numbers (RRNs) by Approximation but being throw-away distance from the same.

In continuation of Sections 4.8, 4.11 and 4.12, we illustrate here, that by approximation activities, RRNs could be obtained. For good deal of knowledge on approximation one may refer Powell's (1988 and 1992), at least his approaches to ' Piecewise polynomial approximations'.

During senior author's doctoral project, it was desired to obtain tri-quantiles of TRI-VAR probability distribution for trip-lets of correlated-variables , including those with zero correlation. All such quantile-trios that could approximately approach pre-fixed probability level, were to be enumerated meeting the set target to remain of near to the it (the target). Also those to remain, within prescribed margin error say é, a very small number say 0.0000001. For the same suggested for convergence algorithm was of repeated- halving. Such suggestion was then given by Prof. Dr. G.P. Bhattacharji , I.I.T, Kharagpur, one of mentors of the said author. Later on, rather only recently the same was found by the author in *Durga-sapt-sati, at its Sri-devya-atherva shirsam* given herein below:

वेदोऽहमवेदोऽहम् । विद्याऽहमविद्याऽहम् ।

अजोऽहमनजोऽहम् । अर्धश्चोऽर्धं च तिर्यक्चाहम् ॥४॥

Vedo-aham avedo-aham , vidya-aham avidya-aham :

Aja-aham anaja-aham, ardhashyo-cho-ardhyan cha

tiryak-cha-aham.||4||.

Import: The Godess utters ' I am in Veda also in what is not in Veda, I exist in knowledge as well as, whatever is not within the ambit of knowledge, I exist in whatever is born, also in whatever is unborn: further I am approachable in convex and non-convex and ' amongst curve-linearity in their every knock and corners' when approached by repeated- halving technique .

The contemplated algorithm is for approaching the numeric- target by small step for each one iteratively. Having approached so, make shift to next nearby numeric entity, to achieve the same in sequence. Thus a sequence of approximants gets created. That the set objective of approximation , is being squarely met with.

5.2: Emergence of gap to be taken as 'error' the sources of fetching (RRNs) :

Now each approximants though very near its target, leaves some gap between former to latter. That is usually being called errors, which naturally and necessarily have to be haphazard. It is those that provide real random numbers (RRNs). Such creation is merely incidental not objectively sought after. Usually no one bothers, or is required to control on its density, magnitude or direction. It is just like thrown away into dust-bin, just a through-away distance from fixed approximation target.

5.3: About mitigation of century old 'philosophical paradox':

Thus it is being inferred that such approximation venture usually yield some minor gap say 'error'. It is such 'error' that is having haphazard or random nature. The same is quite incidental not created with any prefixed objective. Thus it is taken to mitigate a century old persisting 'philosophical paradox' , implying the term 'random' must not have been created while meeting some prefixed objective or purpose.

5.4: Steps of Algorithm: Examples and Graphic Illustrations

Example 1:

It is to recollect that a FORTRAN programme was finalised on 15th. November 1984 to obtain tri-quantile trios for any probability level for given set of correlations in-between three variables. A small sample Table is being presented to display the same:

TABLE No. 5.5.1

The algorithm for approximating step:

N = 30, The Target Scalar = 0.9025: Approximants: Errors = (Target- Approx.);

RRN-1 =ABS(Error) *10^9: RNN-2 = 10000+[(Error)×10^9]

Sl. No.	Target	Approximant	Error	RRN-1	RNN-2
1	0.9025	0.902493942	0.000006048	6058	16058
2	0.9025	0.902498548	0.000001452	1452	11452
3	0.9025	0.902501153	- 0.000001153	1153	8847
4	0.9025	0.902499431	0.000000569	569	10569
5	0.9025	0.902496035	0.000003965	3965	13965
6	0.9025	0.902505592	-0.000005592	5592	4408
7	0.9025	0.902497548	0.000002452	2452	12452
8	0.9025	0.902503636	-0.000003636	3636	6364
9	0.9025	0.902492802	0.000007198	7198	17198
10	0.9025	0.902497371	0.000002629	2629	12629
11	0.9025	0.902503212	-0.000003212	3212	6788
12	0.9025	0.902495795	0.000004025	4025	14025
13	0.9025	0.902500251	-0.000000251	251	9749
14	0.9025	0.902494931	0.000005069	5069	15069
15	0.9025	0.902496045	0.000003955	3955	13955
16	0.9025	0.902500369	-0.000000369	369	9631
17	0.9025	0.902503467	-0.000003467	3467	6533

18	0.9025	0.902509951	-0.000009951	9951	49
19	0.9025	0.902501429	-0.000001429	1429	8571
20	0.9025	0.902498351	0.000001649	1649	11649
21	0.9025	0.902494180	0.000005820	5820	15820
22	0.9025	0.902503082	-0.000003082	3082	6918
23	0.9025	0.902506833	-0.000006833	6833	3167
24	0.9025	0.902494257	0.000005743	5743	15743
25	0.9025	0.902496678	0.000006678	6678	16678
26	0.9025	0.902498599	0.000001401	1401	11401
27	0.9025	0.902500254	-0.000000254	254	9746
28	0.9025	0.902497043	0.000002957	2957	12957
29	0.9025	0.902499039	0.000000961	961	1961
30	0.9025	0.902491653	0.000008347	8347	18347

The fifth and sixth columns yield four and five digits RRN-1 and RNN-2,s respectively. Thus a century old search for the same is over, without resort to PRNs. One may take Abs (the absolute) for RNN-1 and (+10000) for RNN-2 to avoid negative RRNs, as in such form they are required to be of use. The RNN-2 may be preferred as they reflect wider scatter as reflected by original- signed errors.

This follows the MATLAB graphs. The first one portrays the random scatter of the RRNs as obtained in Table No.5.5.1, Col. 5 and then of Col.6. The corresponding magnitude is on right side of scatters graph.

Note:- None, should try to join such scatter random points by joining them by lines, linear or curvilinear, as each of them are separable as made clear also at Section 7.4 of Chapter 7 and Chapter 8. Joining such points carries meaning only when each point of joining lines has to have some realistic connotation for its value of y dependent on corresponding value of x. Every entity of RRNs being separable and independent loses such feature.

The first graph portrays the random SCATTER on LHS graph for RRN-1s and RRN-2s their corresponding MAGNITUDES on RHS graph

RRN-1

RANDOM- SCATTERS

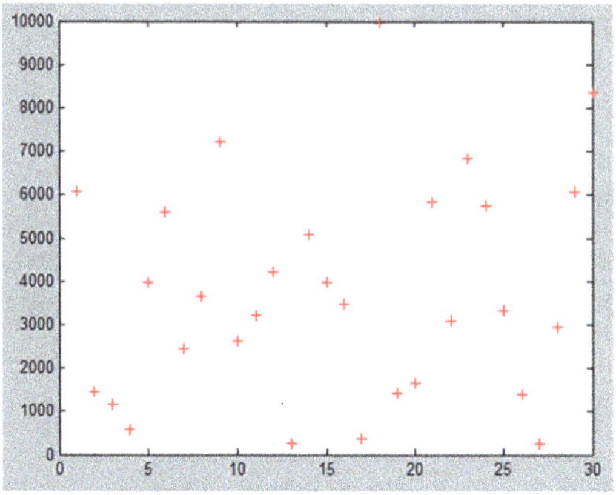

RANDOM-MAGNITUDES

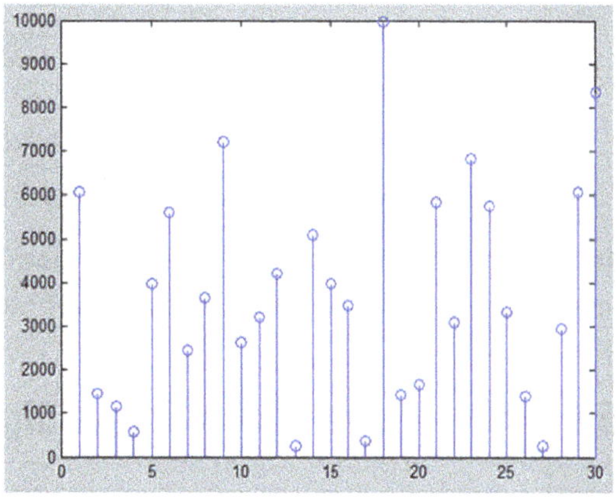

RRN-2

RANDOM- SCATTERS

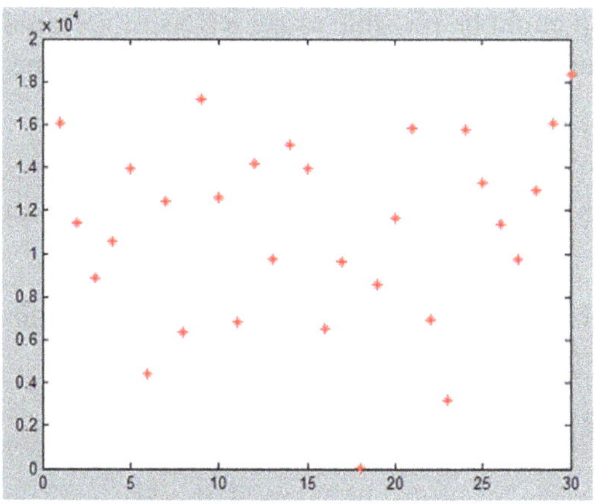

RANDOM-MAGNITUDES

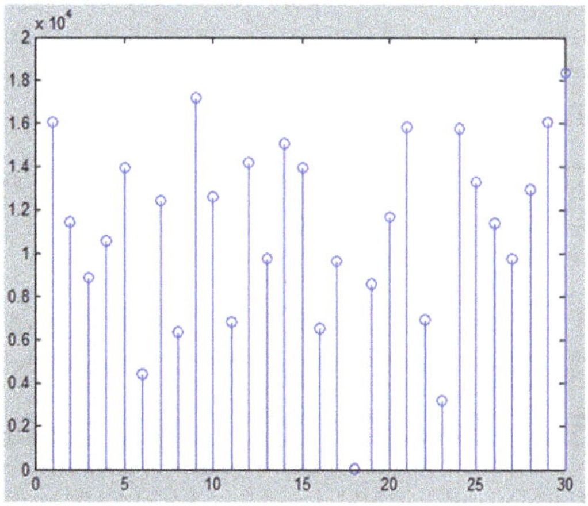

The second graph portrays the magnitude of obtained for RRN-2s.

The graph here attachés magnitude of each such RRNs separately, quite independent of the magnitude of any other. The statement made at Section 8.4 about feature of (Hilbert Space) being separable, gets amply demonstrated here.

As such RANDOM features , now brought under the ambit of 'OBSERVABLES, no THEOREM is REQUIRED to PROVE them. This is by virtue of Kapil's SAKHYAKARIKA-NO. // 4//, being quoted here under, ones again:

दछटमनुमानमास्वचनं च सवेप्रनाणसिडत्वा

चिधिधं प्रमाणभिष्धं व्रवेथसिद्धिः चमाणाद्धि ॥४॥

Thirsta-anumanam-apt-bacha-nancha sarva-pramana-sidhwa-twata;

Trivi-dham pramaanam-istam premeya-sidhhyah pramana-adhi //4//

Import: While talking about proof , the same gets implied with direct observations, (listening, smelling, taste-ting and with touching) to be added or accumulated with writings and listening of authorities.

5.5: On requirement of larger digit- RRNs for encryption:

For the same one may assemble two and three sequences of such numbers. That may yield $^{N}P_{n}$ such numbers. For N=30 and n=2, it is to be 30×29= 870, RRNs.

5.6: Notable aspects indicated:

It must be noted that by the process of 5.5 here, it could be possible to get RRNs of required size for almost most of the users, for their varied uses on different occasion, purpose and time. That indeed is great achievement.

Though, we have achieved the desired results and have completed the purpose of this Chapter but suggestion came that one such example may not be sufficient to convince most readers and users.

Therefore some more examples from own published monograph Das(2015) but its non-published version is being presented as addendum the this Chapter.

5.7: Examples 2, 3 & 4(to conform the process):

Here we take, RNN-1 = Abs(Error) $\times 10^9$ and RNN-2= [(Error)$\times 10^9$]+10^8 ,so that we may get them to be a whole numbers.

TABLE NO. 5.6.1

TABLE OF ERRORS and THEIR CONVERSION to RRNs

N	Derived Error	RRN-1	RRN-2
1	0.0003218651	3218651	13218651
2	0.0003099442	3099442	13099442
3	-0.0003337860	3337860	6662140
4	0.0004768372	4768372	14768372
5	0.0005185604	5185604	15185604
6	-0.0002205372	2205372	7794628
7	0.0001788139	1788139	11788139
8	-0.0003576279	3576279	6423721
9	-0.0000417232	417232	9582768
10	0.0009298325	9298325	19289325
11	0.0007033348	7033348	17033348
12	-0.0003695488	3695488	6304512
13	-0.0001132488	1132488	8867512
14	-0.0001430512	1430512	8569488
15	0.0009417534	9417534	19417534
16	-0.0008761883	8761883	1238117
17	-0.0001966953	1966953	8033047
18	-0.0000357628	357628	9642372
19	-0.0000834465	834465	9165535
20	-0.0007808209	7808209	2191791
21	0.0009119511	9119511	19119511
22	0.0004172325	4172325	14172325

RRN-1

RANDOM- SCATTERS

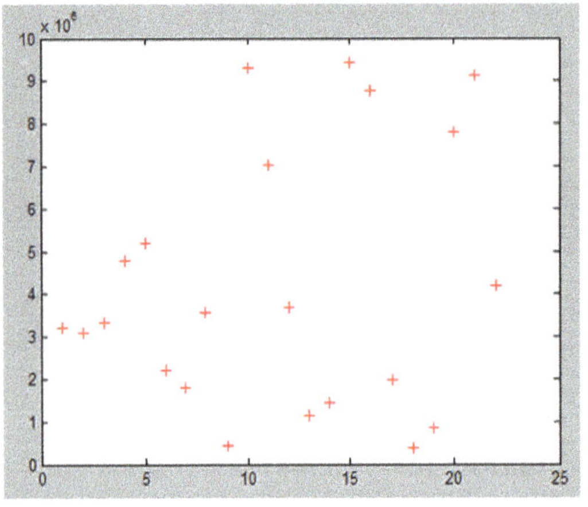

RANDOM-MAGNITUDES

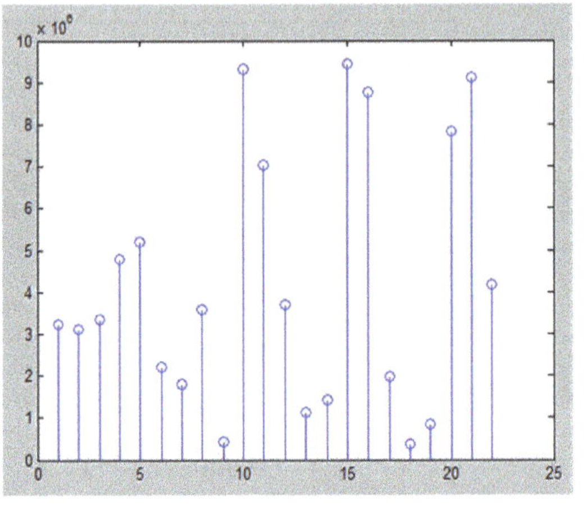

RRN-2

RANDOM- SCATTERS

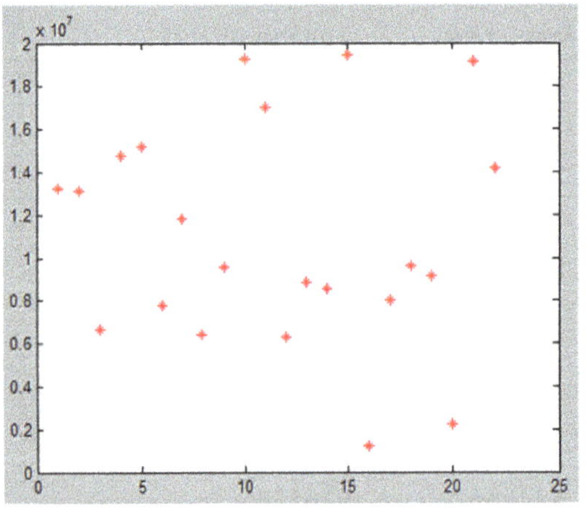

RANDOM-MAGNITUDES

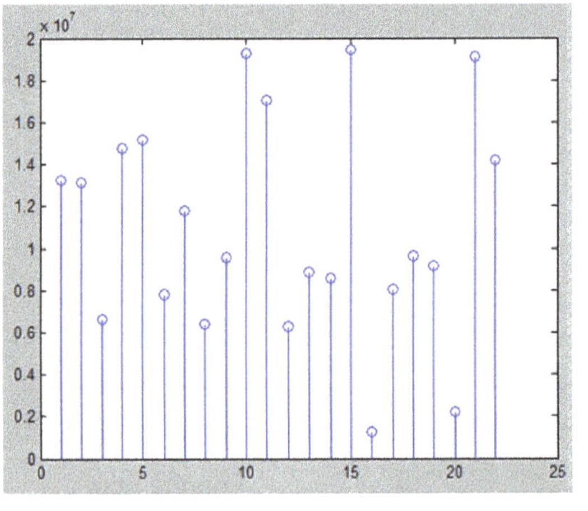

All the statements made for the RRNs in TABLE NO. 5.5.1, appears hold good for this Table also.

Example 3:

Like previous example we take RNN-1 = Abs(Error) $\times 10^9$ and

RNN-2 = [(Error)$\times 10^9$] + 10^7

TABLE 5.6.2

N	Derived Error	RRN-1	RRN-2
1	-0.0003159046	3159046	6840954
2	0.0006198883	6198883	16198883
3	0.0004172325	4172325	14172325
4	0.0008761883	8761883	18761885
5	0.0009655952	9655952	19655952
6	-0.0003278256	3278256	6721744
7	-0.0005543232	5543232	4456768
8	-0.0003099442	3099442	6900558
9	0.0002026558	2026558	12026558
10	-0.0008821487	8821487	1178513
11	-0.0002026558	2026558	7973442
12	-0.0002503395	2503395	7496605

RRN-1

RANDOM- SCATTERS

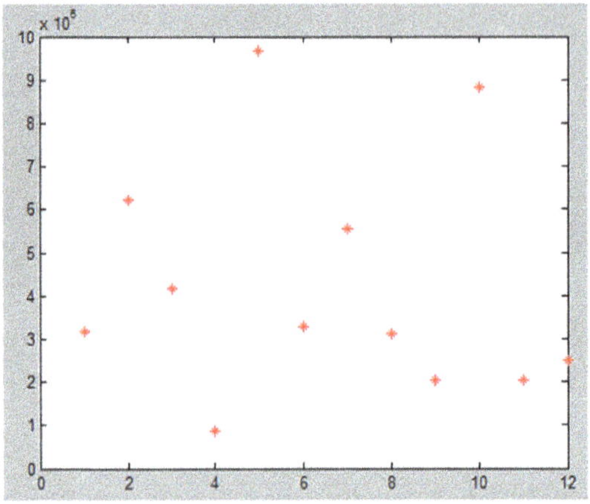

RANDOM-MAGNITUDES

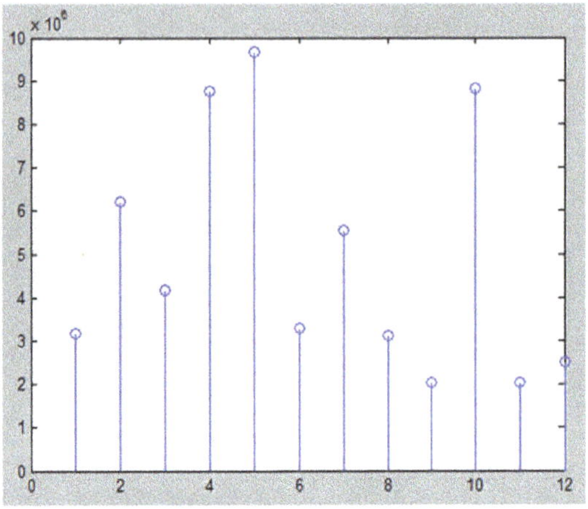

RRN-2

RANDOM- SCATTERS

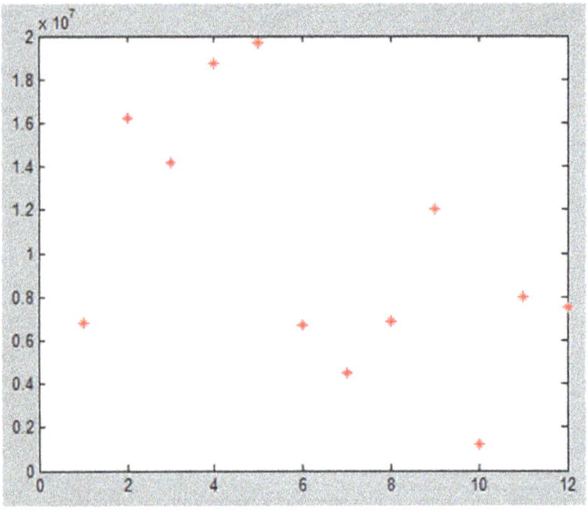

RANDOM-MAGNITUDES

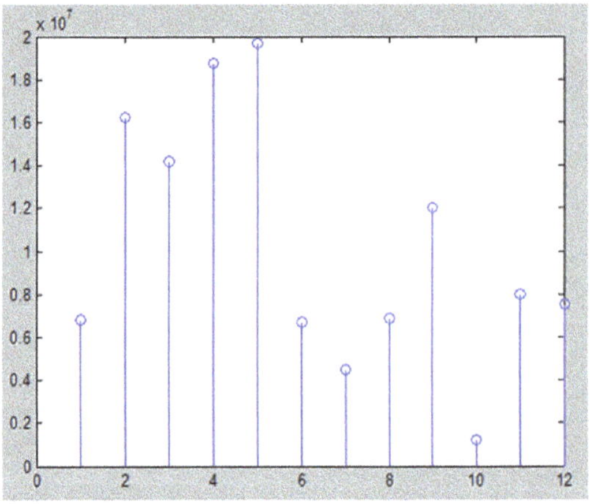

Example 4:

TABLE 5.6.3

N	Derived Error	RRN-1	RRN-2
1	-0.0002443791	2443791	7556209
2	-0.0000298023	298023	9701977
3	0.0005662442	5662442	15662442
4	-0.0004649163	4649163	5350837
5	0.0002980232	2980232	12980232
6	0.0007569790	7569790	17569790
7	-0.0008583069	8583069	1416931
8	-0.0001192093	1192093	9807907
9	0.0008583069	8583069	18583069
10	0.0003099442	3099442	13099442
11	-0.0007688999	7688999	2311001
12	-0.0005960465	5960465	4039535
13	-0.0001907349	1907349	8092651
14	-0.0005722046	5722046	4277954
15	0.0005245209	5245209	15245209
16	-0.0005006790	5006790	4993210
17	-0.0005304814	5304814	4695186
18	0.0006020069	6020069	16020069
19	-0.0003993511	3993511	6006489
20	0.0003933907	3933907	13933907

RRN-1

RANDOM- SCATTERS

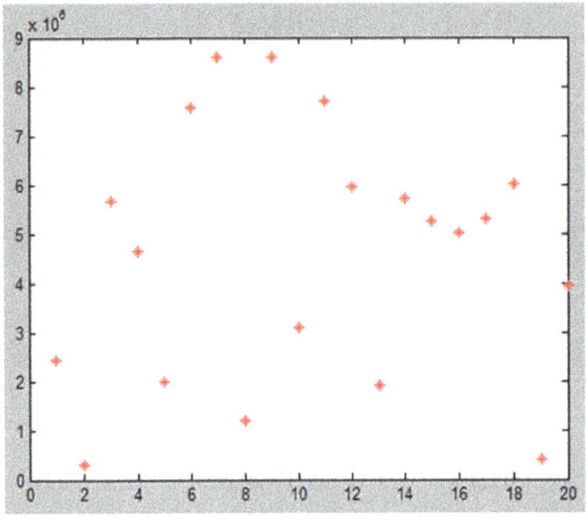

RANDOM-MAGNITUDES

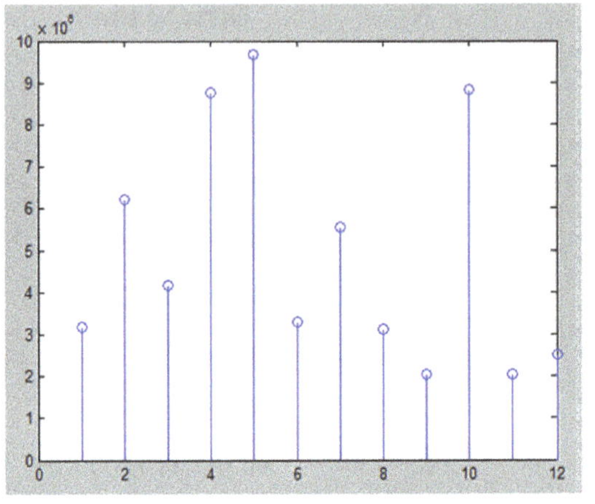

RRN-2

RANDOM- SCATTERS

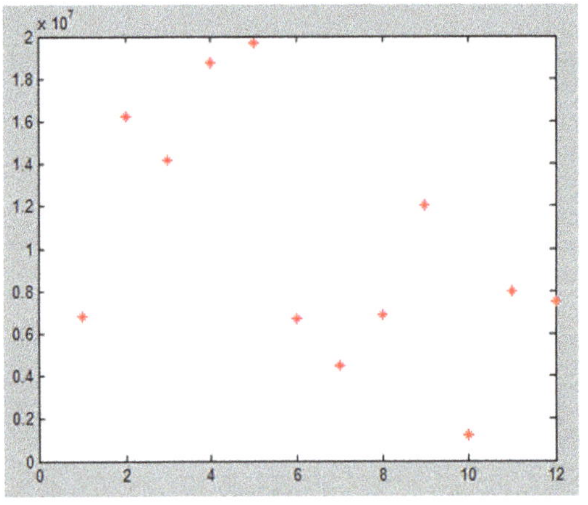

RANDOM-MAGNITUDES

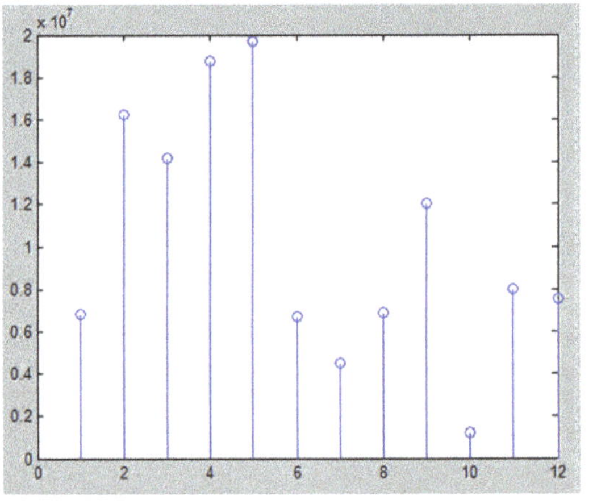

5.8: Concluding Remarks for the Chapter:

We now have been able to get reasonable number of RRNs by approximation algorithm, along with their graphic MATLAB illustrations. Those authenticate their being RRNs.

We have large set of such computational 'Errors': those may be made available on demand by genuine buyer of the book by converting them to RNNs. Those may have to be chucked off from our stock so that the same is not to given to some users.

References

1. Das, N.C.: (1984): 'TRIVAR' , An Unpublished Work, of reasonably large size available for users on demand. As, its publication, shall make it unusable for encryption for any other user.

2. Das, N.C.: (2015): Decision Processes by Using Bivariate Quantile Pairs, Springer India New Delhi: (from its pre-published versions).

3. Maharishi Kapil; (2007): Sankyakarika of Iswar krisna by R.S. Tripathi,(1970), Chowkhamba Krtshnadas Academy , Varanasi, U.P. , India

4. Powell, M.J.D. (1988) and (1992): Approximation theory and methods and possibly " The theory of radial basis functions in 1990".

5. Rishi Markandey' (2024): Durga-sapt-sati: Sri-Devya-atharva shirsam||4||, Gita Press, Gorakhpur, UP, India.

CHAPTER 6

Obtaining RRNs Through Filtering

6.1: Search for RRNs from residues of filtering:

Search of RRNs in residues and/or remains (as though on sieves) of filtering but not the filtered produce.

In preceding Chapter, it could be seen that RRNs are obtainable as residuals of approximation. For the present chapter it is required to be investigated as to whether it is possible to find existence of RRNs while data being filtered.

Therefore let us first try to be clear as to what filtering means here, in what way and where filtering is implicitly required. In fact filtering presumes that data being obtained are polluted and/or mix-up of desirable with some undesirable elements and that makes filtering essential. In fact there are many materials and aspects like mining ores, crude petrol and waters of the rivers, food-grains by its husks and even sounds that we listen are disturbed with noise, those essentially require filtering. Also such undesirable elements vary randomly in time and space. Hence the discipline like RANDOM PROCESSES, implicitly need measure of uncertainty, called probability. That necessitates some mathematical functions along with its fitting methods, to be used as filtering device based on sound logic and/or reasoning processes.

Difference between approximation and filtering lies in understanding that while in former, target- scalar remains fixed and its remnants, a random: while in filtering both , crude measurement and/ or observation and thus, their 'remains' are random. In case filtering device happens to be poor, their remnants implying RRNs will be of larger magnitude. It is suggested therefore to improve the filtering device i.e. to change mathematical function and/ or better estimation process, say improved Kalman's filter. Some such methods, those have been suggested by Das and Das (2021) and Das and Ranjan (2024). Where, they have presented, new simple methods respectively for Linear and Non-linear function fitting to such data-set as filtering devices.

Therefore as caveat and caution, it is advisable to suggest that (filtering apparatus) fitting function and its estimation process must be efficient and near perfect: then only the remains may possibly be called genuine error , otherwise not the same.

It may sometimes be necessary to conduct Neyman's Chi-Sqr. Test for its (error) non-significance, as we have done in Das and Das (2021), to change the filtering device i.e. the mathematical function and/ or estimation processes.

6.2: Filtering Process, implying fitting Process:

Virtually fitting process implies estimation process, which in effect means estimation. That to mean estimation of parameters of filtering device, which usually is mathematical function fitted to observed or measured data-set. The same as Ludeman(2010) puts as:

He identifies three major types of estimation problems. For the same he creates two time indices j and k, then classifies undernoted those three types as:

1) If $k > j$, the problem is of *prediction.*
2) If $k = j$, the problem is of *filtering.*
3) If $k < j$, the problem is of *smoothing or of interpolation.*

Whatever be the problem among those three, he chooses any mathematical function f(xj) to approximate the observation or measurement believably filtering of some crude. He is then required to define an estimation error as:

$$\acute{e}(x_j) = x_j - f(x_j)$$

The same is required RRN, as it happens to be the residue, not the objective of the filtering processes, irrespective of the three estimation problem, so called Filtering Processes.

6.3: Example 1, followed by Examples 2,3 and 4:

The data for the Example pertains to Ludeman(op cit) but its filtering has been done by Das & Das (2021) that being SLIM and GUDFIT process, displayed at Section 8.2.7 and Table 8.2.9 of the latter book referred. There the filtering device chosen and fitted was a linear function:

$y(j) = 7.910237 + 6.848997 \times x(j)$,

Columns of the Table 6.3.1:

Column :1 The index j,

Column: 2 *x(j)*, the observation or measurement requiring filtering,

Column: 3 *f(x(j))*, the filtering device ,to be a suitable mathematical function, estimated from *x(j)*.

Column: 4 diff *(x(j)-f(x(j))*, say *diff(j)*

Column: 5 RRN-1(j) = abs(diff(j))×r, r is a positive scalar multiplayer to change it, to a whole number, as and when necessary.

Let, r = 10^6 = 10,00,000:

Also for Column:6; RRN-2(j) = 10000000+(diff(j)×r) as diff(j) has to have both positive (+) and negative (-) signs.

The same table is now Table 6.3.1:

TABLE 6.3.1

Col. No.	1	2	3	4	5	6
	j	x(j)	f(x(j))	diff(j)	RRN-1(j)	RRN-2(j)
	1	12.9667	14.75923	- 1.792534	1792534	8207416
	2	22.8005	21.60823	1.192270	1192270	11192270
	3	31.6740	28.45723	3.216774	3216774	13216774
	4	25.7334	35.30622	- 9.56282	9562820	437180
	5	49.1115	42.15522	6.956290	6956290	16956290
	6	44.7559	49.00422	-4.248314	4248314	5751686
	7	58.4817	55.85322	2.628487	2628487	12628487
	8	64.3720	62.70221	1.669792	1669792	11669792
	9	67.7068	69.55121	-1.844406	1844406	8155596
	10	78.1946	76.40021	1.794396	1794396	11794396

RRN-1

RANDOM- SCATTERS

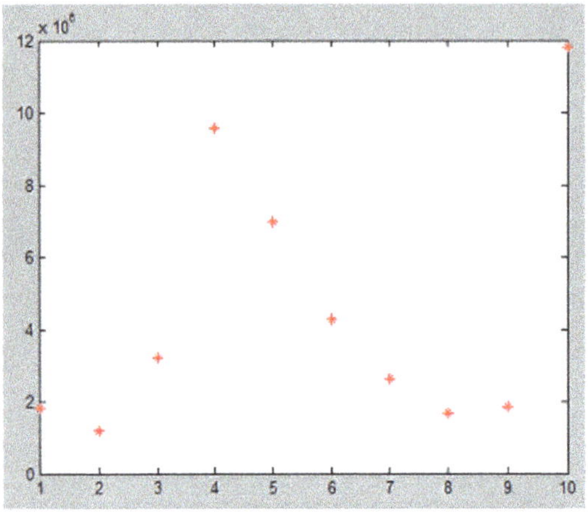

RANDOM-MAGNITUDES

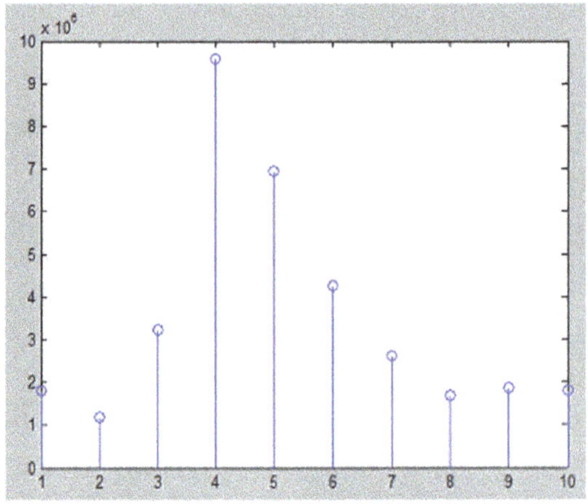

RRN-2

RANDOM- SCATTERS

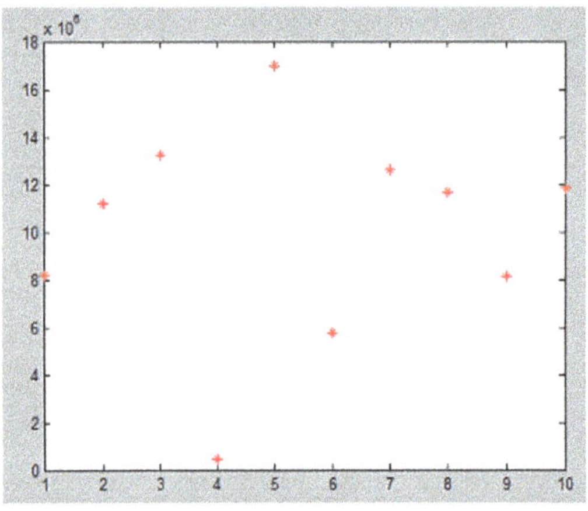

RANDOM-MAGNITUDES

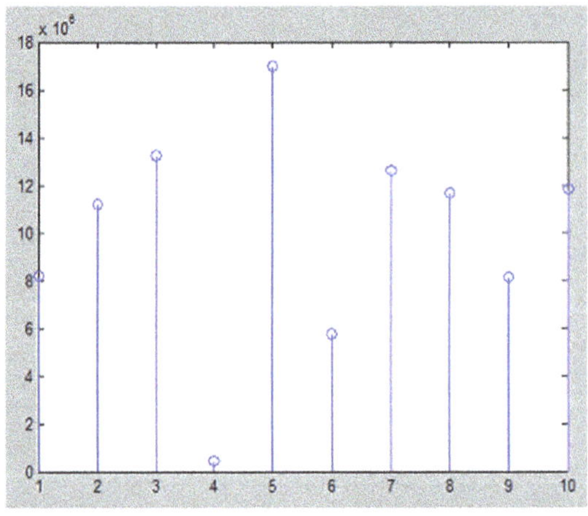

Thus columns 5 and 6 yield satisfactory RRN-1 and RRN-2s.

The example displays that even if the observations or measurements be random variable the random terms can be isolated as RRNs. Once again we assert that random number so obtained is random and nothing else without any test for the same. Thus the each of the two RRNs has to have different values and graphs but both are random, to be of advantage for the users.

Example 2: We may derive RRNs from the data-set at Table 2.5.2 of Das and Das (2021), pp 15. There the fitted filtering device was:

y-pred (i) = 22.5 + 4.5× y(i)

Here the, diff[y(i) - y-pred(i)] = error(i);

Abs[error(i)]×10^5 = RRN-1(i);

and

RRN-2(i) =[error(i)×10^5]+ 1200000.

TABLE 6.3.2

(i)	y(i)	y-pred(i)	diff[y(i) – y-pred(i)	RRN-1(i)	RRN-2(i)
1	69	63.0	6.0	600000	1800000
2	76	76.5	-0.5	50000	1150000
3	52	49.5	2.5	250000	1450000
4	56	67.5	-11.5	1150000	50000
5	57	63.0	-6.0	600000	600000
6	77	67.5	9.5	950000	2150000
7	58	54.0	4.0	400000	1600000
8	55	58.5	-3.5	350000	850000
9	67	76.5	10.5	1050000	2250000
10	53	49.5	3.5	350000	1550000
11	72	72.0	0.0	000000	1200000
12	64	58.5	5.5	550000	1750000

RRN-1

RANDOM- SCATTERS

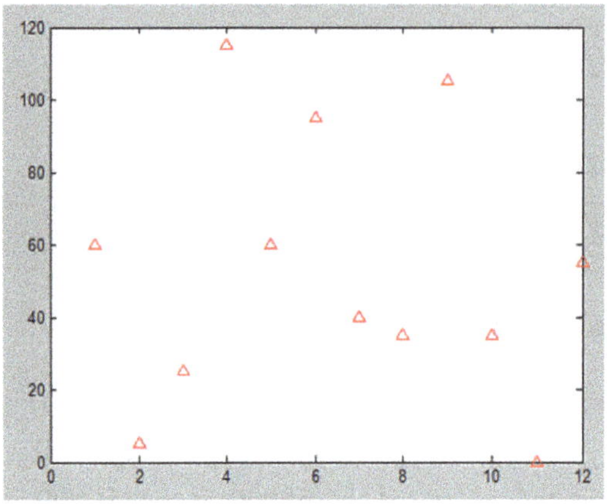

RANDOM-MAGNITUDES

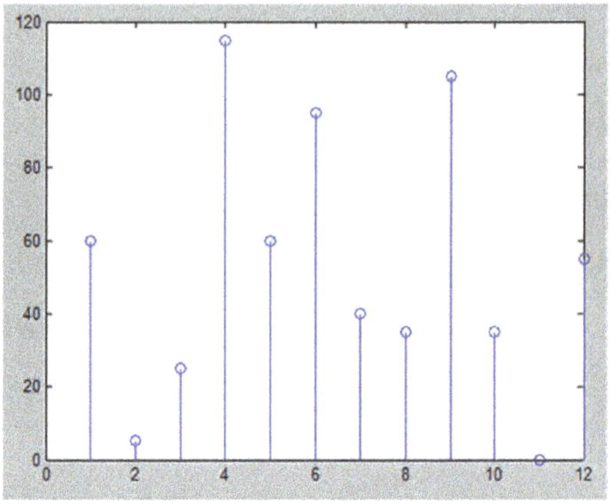

RRN-2

RANDOM- SCATTERS

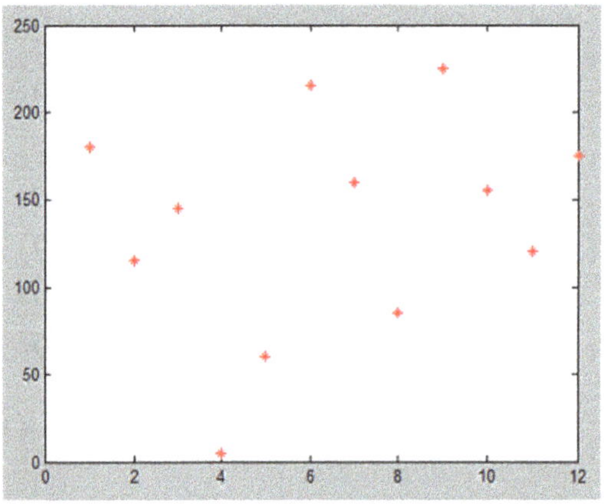

RANDOM-MAGNITUDES

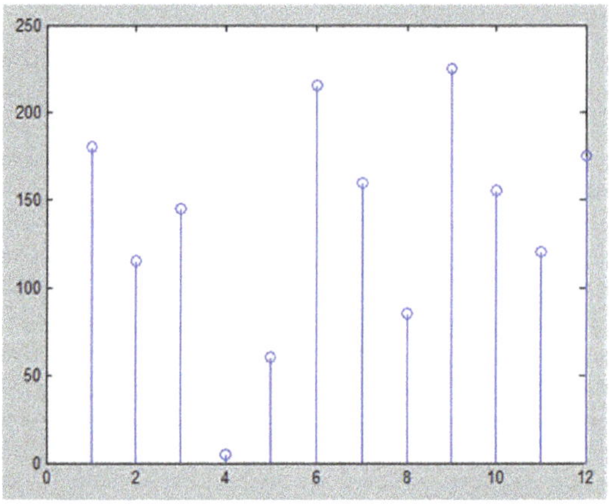

The linear filtering function is not good enough hence the RRNs deviations are very large but that's very natural: nevertheless they are as much random, as could be.

Example 3: To obtain RRNs from Koutsoyiannis's (1996) data-set.

Filtering device is fitted linear equation to given data-set: y(i),s in the Table hereunder:

$$y\text{-pred}(i) = 31.82547 + 0.7091563 \times y(i).$$

RRN-1(i) = Abs[error(i)]×(10^5)

and

RRN-2(i) = [error(i)×(10^5)] +100000

TABLE 6.3.3

(i)	y(i)	y-pred(i)	error (i)	RRN-1(i)	RRN-2(i)
1	282.3	287.0508	-4.7508	47508	52492
2	291.1	294.2842	-3.1842	31842	68158
3	312.3	311.7295	0.5705	5705	105705
4	326.5	325.7708	0.7292	7292	107292
5	336.6	337.1173	-0.5173	5173	105173
6	356.6	357.1155	-0.5155	5155	105155
7	376.6	374.7025	1.8975	18975	118975
8	402.9	397.3956	5.5044	55044	155044
9	434.7	427.1092	7.5908	75908	175908
10	468.3	466.6092	1.6908	16908	116908
11	494.3	499.0886	-4.7886	47886	52114
12	538.9	543.1271	-4.2271	42271	57729

RRN-1

RANDOM- SCATTERS

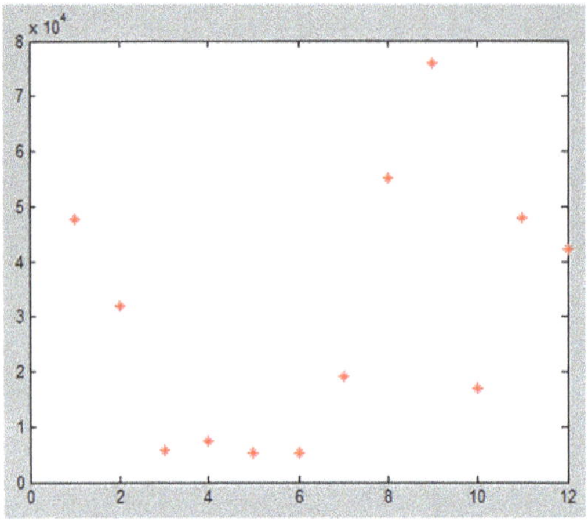

RANDOM-MAGNITUDES

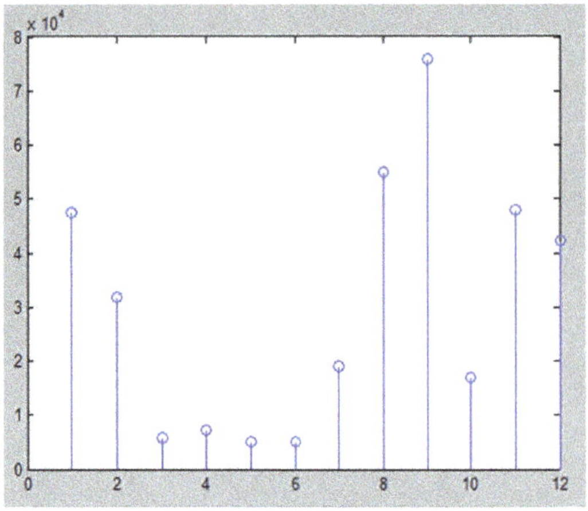

RRN-2

RANDOM- SCATTERS

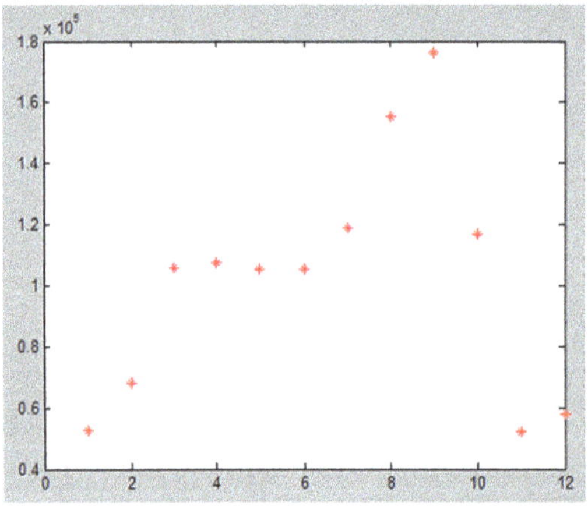

RANDOM-MAGNITUDES

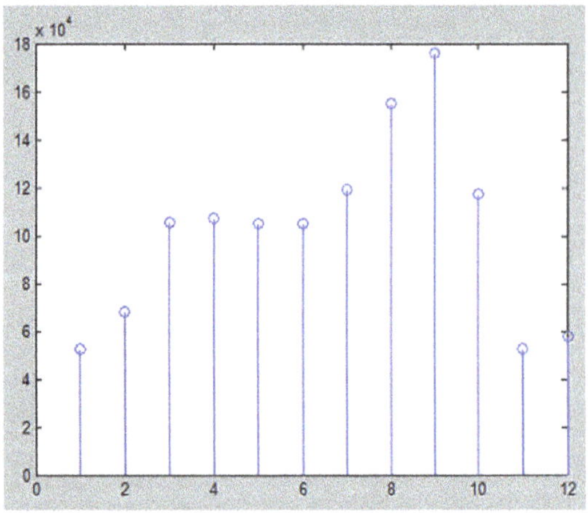

The fifth and sixth Columns present respectively RRN-1 and RRN-2, those appear satisfactory.

Example 4: Obtaining RRNs from Levin et al (2013) data-set.

Filtering device is a Linear Model:

$$y(i)^\wedge = -2.972119 + 0.7122942 \times y(i) ,$$

obtained by fitting to the data-set y(i), the observed data set needing filtering.

The error(i) = y(i) – y(i)^ ,

RRN-1(i) = Abs(error(i))×10^6 and RR N-2(i)= [error(i)×10^6]+1000000

TABLE 6.3.4

(i)	y(i)	y(i)^	error (i)	RRN-1(i)	RRN-2(i)
1	6.2	6.287706	-0.087706	87706	912294
2	8.6	8.424588	0.175412	175412	1175412
3	7.2	7.0	0.2	200000	1200000
4	4.5	4.86 3117	-0.363117	363117	636883
5	9.0	9.136882	-.0.136882	136882	863118
6	3.5	3.438529	0.061471	61471	1061471
7	6.5	6.287706	0.212294	212294	1212294
8	9.3	9.136882	0.163118	163118	1163118
9	9.5	9.847976	-0.347976	347976	652024
10	5.7	5.575412	0.124588	124588	1124588

RRN-1

RANDOM- SCATTERS

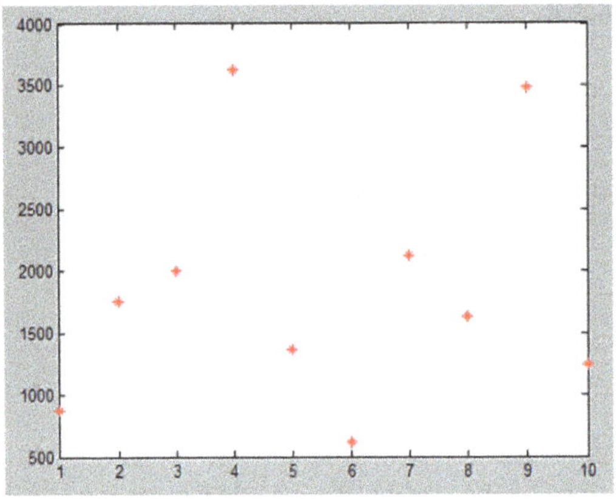

RANDOM-MAGNITUDES

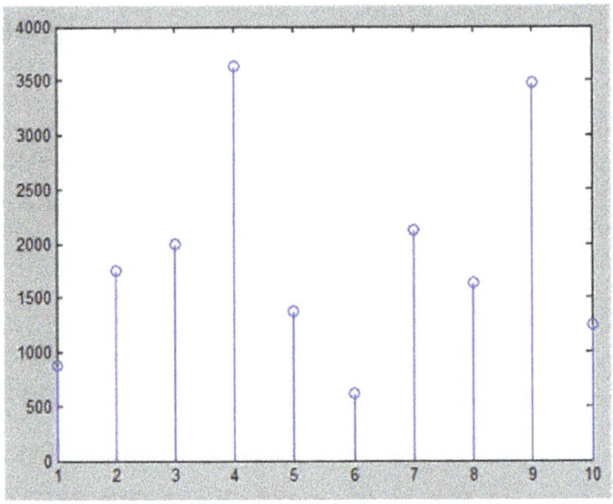

RRN-2

RANDOM- SCATTERS

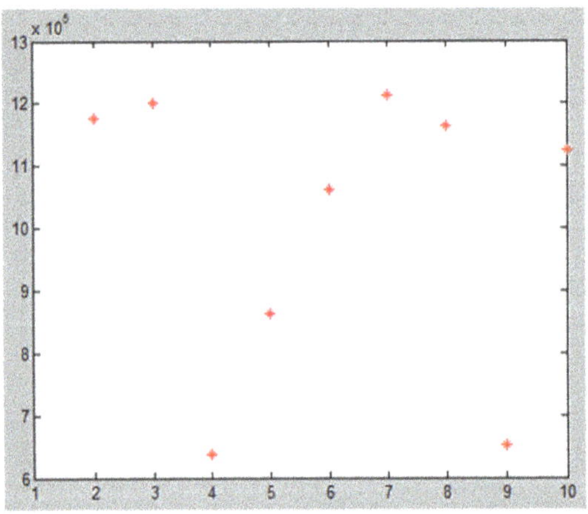

RANDOM-MAGNITUDES

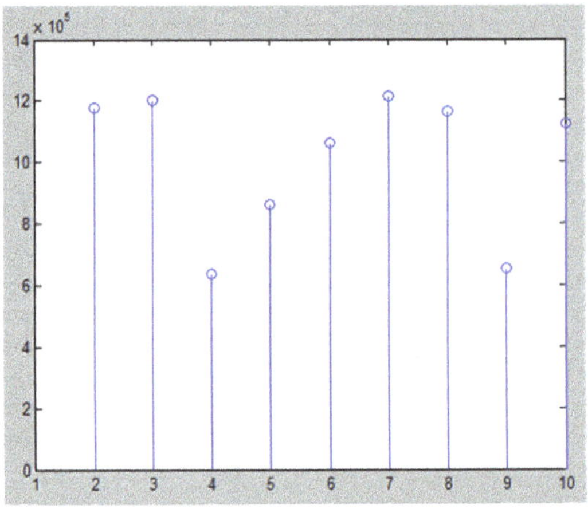

Though four examples we could demonstrate that the remnants of filtering may yield RRNs.

6.4: Chapter concluding remarks:

To increase digital length of RRNs. , one may choose randomly one from Chapter 6 and another from Chapter 5, in sequence and or its reverse sequences with caution that any specific digit do not often gets repeated more than twice , though there is no ethical binding for the same.

For more such examples reader may please refer Das and Das (2021) or any other text on Regression Models. In fact the same text have several examples, those may be of use for such applications.

In next chapter we may envisage to get RRNs from Bhakaracharya' well known orthogonal -trios, those are 3, 4 and 5 and its extensions by using Diophantine Equations.

References

1. Das, N.C.and Das, Mohit R.:(2021) : Data Science for Prediction, Shroff Publishers & Distributors Pvt. Ltd. Sanpada , Navi-Mumbai 400705.

2. Koutsoyiannis, A. (19960: Theory of Econometrics: Macmillan Press Ltd, Hamshire, UK

3. Levin, R.I., Rubin, D.S., Rastogi, S. and Siddique , M. H. (2013): Statistics for Management, 7th. Edn. ,Pearson Edn. New Delhi

4. Ludeman, L. C. :(2010): Random Processes, Filtering, Estimation and Detection., Wiley India, New Delhi.

CHAPTER 7

Obtaining RRNs through Bhaskaracharyar's Trios
and
Quadratic Diophantine Equations

7.1: Introduction:

While on laptop for this Chapter, it came to my memory as to how content of this Chapter could be associated with generation of RRNs. It was my long association with fertilizer trials at farmers' fields in different districts of Bihar, I mean Bihar, before its partition as Bihar (1953-57) and in second phase(1984-93) in Jharkhand, a separate state carved out of then-Bihar.

We use to select village and then farmers' fields with Random Number Table, that by Kendall and Babington Smith (1938-9). A copy of its initial part (with its only 100 numbers) is being placed at Appendix 4.

Immediately thereafter we use to take up Bhaskaracharya's (1151) singleton digit-trios : 3, 4 and 5, to form a Right Angle Triangle at a point in selected plot, then to form a rectangular plot. Such practice was in use since long, perhaps traditionally. That was how Random Number and Bhaskaracharya's Trios(BTs) could incidentally find association by us, Das (1953-59) and 1984 - 93). We may call such space generated by BTs as LIL-BHAS orthogonal space. During my earlier part, I (the Senior Author) remained in contact with Dr. Frank Yates at Field Experimental Service(FES), then at Patna.

How it can find a niche here? Yes, both RRNs and BTs are real numbers, being elements of Number Theory, by virtue of that both can very well carve niche here.

7.2: Bhaskaracharya's - Trios(BTs) and its Pythagorean Property:

Clearly BTs, such feature is also Pythagorean (orthogonal) property to form a right angle triangle, at Farmer's field. BTs have special property to span or generate finite/ to < (that is less than countable- infinite integer-number) space of the size as large as required.

1). $(3^2 + 4^2) = 5^2$, is the first Pythagorean - identity at the same time, the first Quadratic Diophantine Equation, which is due to Bhaskaracharya(op cit) through his famous book LILAVATI. Refer Knuth (1997) also. It is also the first Diophantine Equation as being called in Number Theory. Refer Fermat's last theorem, there in. other sentence

Obviously this is a mathematical truth for any and every multiplication, naturally for all the undernoted Diophantine Equations, which are being enumerated below, taken from , In fact it was Fermat's Last Theorem that yielded Diophantine Equations of triples, the list which has been given by Herkommer(1999) as being presented herein below:

7.3: Herkommer's List of Diophantine Equations.

2). $5^2 + 12^2 = 13^2$:	3). $7^2 + 24^2 = 25^2$:	4). $8^2 + 15^2 = 17^2$:
5). $9^2 + 40^2 = 41^2$:	6). $11^2 + 60^2 = 61^2$:	7). $12^2 + 35^2 = 37^2$:
8). $13^2 + 84^2 = 85^2$:	9). $15^2 + 112^2 = 113^2$:	10). $16^2 + 63^2 = 65^2$:
11). $17^2 + 144^2 = 145^2$:	12). $20^2 + 21^2 = 29^2$:	13). $28^2 + 45^2 = 53^2$:
14). $33^2 + 56^2 = 65^2$:	15). $36^2 + 77^2 = 85^2$:	16). $39 + 80^2 = 89^2$:
17). $48^2 + 55^2 = 73^2$:	18). $65^2 + 72^2 = 97^2$:	19). $17^2 + 144^2 = 145^2$.

7.4: Deriving RRNs from Bhaskaracharya's (BT) to Herkommer's List.

Each of them can be multiplied by any real number to enlarge or reduce its size and/ or their spaces. It is to recollect that the space of real number happens to be much larger that of large prime numbers on which PRNs are required to be based. Also of their arbitrary assemblages as pairs and/or trios may be able to create real random numbers (RRNs) of various digital length. As for example say the pairing of No. 5) and No. 16) are able to create 412 and 892 = 1681and7921 respectably and their inverse arrangement 7921, 1681 as eight digit RRNs. Similarly through their trios, still longer digit sizes, at the same time larger number of RRNs.

Also their number of pairs by random permutation, may themselves be $19 \times 18 = 342$ and trios to be $= 19 \times 18 \times 17 = 5814$. Further they would be of different digital lengths, those being realistic demand and requirement of the time and space. Orthogonal space, so generated deserve to be called LILA-BHAS space.

7.5: Role of LILA-BHAS Space to Hilbert's space:

Much more interesting part is required to be played by LILA-BHAS space as of Hilbert Space for RRNs and its scope for application in Chapter 8.

7.6: Caveat and Caution:

Of course it must never been disclosed as to which of the five methods and/or their joint sequence or their discrete amalgam have been used when and where for encryption and/or coding.

References

1. Bhaskarachaya (1151): Lilavati (PATIGANITA) a book on Aryan (Hindu) Mathematics, Edt. By Prof. Ramchandra Pandey, BHU, Chawkhamba Krisnadas Academy, Varanashi, UP,India

2. Herkommer, M.A.:(1999): Number Theory: A programmer's Guide, McGraw-Hill , New York, U.S.

3. Kendall, M.G. and Bobington Smith, B.:(1939): Tables of Random Sampling Numbers, Cambridge Univ. Press , Cambridge, London

4. Knuth, D. E.:(2000): The Art of Computer Programming, Vol.2, Addison Wesley, California, US

CHAPTER 8

On use of Hilbert Space employing Topology to create RRNs

8.1: Introduction:

The Chapter embarks for opening an innovative venture of Hilbert Space to create RRNs for its multiple usages, including encryption. For the same a brief and introductory knowledge on topology is expected of readers. Therefore interested readers are advised to refer, Simmons (1963) or Debnath and Mikosinski (2006). If one is ready to assume that the underlying statements are tangible, then he/she may proceed without the same.

Simmons wherein stated 'The reader may wonder why we restrict our attention to complex spaces. Why not consider real spaces as well ?'

The same, we have already done in previous Chapter 7, and found number of tangible solutions. But Simmons argues, "we could easily do so, and many writers adopt this approach. There are few places in this Chapter, where complex scalars are necessary, but theorems involved are not crucial, and we could get along with real scalars without too much difficulty. It is only the complex case that the theory of operators on Hilbert Space assumes really satisfactory form".

Further he stated that "every polynomials of n-th. degree with complex coefficients has exactly 'n' complex roots (some of which are repeated). For this reasons, we limit ourselves to complex case throughout the rest of the book."

Implying by such consideration one is sure of getting viable root, though for quadratic case both such roots may lie in Argand (complex) space but their inner product being quadratic returns to real space. That we may witness in example cited here in under.

8.2: From Bhaskaracharya's Trios(BT) to Complex (Argand) space to fetch RRNs:

It is in continuation of Section 7.2 of preceding Chapter, we take up as an instance:

$r \times (3^2 + 4^2) = r \times 5^2$, which is true for all real number 'r'.

Now in place of 'r' as multiplier, we choose a complex number 'c' as multiplier and observe the result. Where 'c' is complex number of the type:

1); $c = (d + ie)$, where 'i' is imaginary number = $(-1)^{1/2}$ but 'd' and 'e' are real numbers. That makes 'c' to be a complex number.

2); We also know that every complex number 'c' has a conjugate complex number 'c^ ' which is of the type $c^\wedge = (d - ie)$, as another root.

3); We are also aware of the fact that the product:

$c \times c^\wedge = (d^2 + e^2)$ as $i^2 = -1$.

Taking advantage of these results, while we are required to encrypt at 'source' by multiplying LILA-BHAS's Diophantine Equation's LHS component (3^2+4^2) with c , whereas to decrypt at 'sink' it is needed to multiply with c^ to get the same as product of $(d^2 + e^2)$, those as assumed both d and e are real numbers. Yet the results are to be the members of Hilbert space. The RHS of such equation 5^2 would now be the product of $(d^2 + e^2)$ and 5^2. Such product may be taken as RRN.

Here both 'd' and 'e' are independent real random numbers to be chosen by any of the method mentioned before but their secrecy to be well maintained.

8.3: Examples 1 and 2:

Let such d and e respectively be chosen number 11, and 7. At source it is being sent as multiplied by $(11 + i \times 7)$ but to decode at sink it would be further multiplied with $(11 - i\times7)$, hence their product would be the effective multiplier $d^2 + e^2 = 121+49 = 170$, the result then would be $170\times5^2 = 4250$, decoded at sink .

To observe its impact by change, we take another example, say its No. 11), amongst its nineteen sets.

That is: Example 2:

11): Which is: $17^2 + 144^2 = 145^2$

Let $c = (d + e\,(-1)^{1/2}) = 17 + 144\,(-1)^{1/2}$ and $c^\wedge = 17 - 144(-1)^{1/2}$.

$c\times c^\wedge = [\,(17 + 144(-1)^{1/2} \times (17 - 144(-1)^{1/2})\,] =$

$= 17^2 + 144^2 = 289 + 20736 = 21025$

The method may very well apply to any of the well known Diophantine Equations but their numbers are finite, hence it has to be used with two or three naturally chosen random numbers as its multipliers and/ or other mathematical relations .

8.4: The Separable property of Hilbert Spaces:

It is worth mentioning that each result as obtained by different LILA-BHAS Diophantine Equations is capable of generating Separable Hilbert Space: Refer Debnath and Mikusinaski (2006) for the same. Implying, clearly that each such numbers, obviously RRNs independent and discrete. It is such property of RRNs that reminds of the Kapil's statement made millenniums of years before to be quoted once again:

सप्त्यङ् किल ये अर्थस्तेऽर्थाः, कृतसंन्यास्य षष्टितन्त्राः।

आख्यायिका बृहद्हिता परवद्विवर्जिता तास्य चापि॥ ७२॥

Saptyanga kile ye-artharaste-arthah kritasna shasti-tantrasya:

Aakhya-ye-ka vira-hitah parwad-vivarjitah-asta-chapi ||72||

Import: It is to be beyond any anticipation at the same time neither it is required to be nearer, nor distant from any one. That was how Maharishi Kapil conceived and delivered his concept of 'random' having virtues of having random at the same time Hilbert's independence towards any other members of sequence. Though he meant then, to be acquired by saintly person but the same appears applicable to choice of random number: yes the Real Random Numbers.

This is the sermon to be given through mine (Kapil's shasti-tantra) 'hexagonal system' of this seventy epithets that he propounded through his well known (Sankhaya darshan), the Sankhya Philosphy.

Though for the context here, we are not required to explain his 'hexagonal system'. However implication of such hexagonal system, as being conceived by me (the senior author) has been given here at Appendix 3.

8.5: Using Hilbert Space to get separable RRNs:

Thus Hilbert Space gets new application to be of use for finding Real Random Numbers over and above what we get by other popular existing and now explored devices.

Thus the Number Theory and Topology all together finds new confluence to mingle up for finding RRNs to mitigate almost a century old, philosophical paradox.

8.6: Related Inference:

Although a number is random or not is to be decided of its status in a sequence of other numbers but remains unattached and/or completely independent of other entities of the sequence, rather forms discrete Hilbert space amongst them.

8.7: Maharishi Kapil's concept of 'aakhyaika-vira –'and its congruence/ similitude of 'random':

Thus any test proposed to test its independence or not so happens to be philosophically/ ethically erroneous as the same is not required to be attached or non-attached to any or every members the sequence. As already excerpted earlier the writings of Maharishi Kapil earlier, which may be partially repeated once again:

"aakhyaika vira-hitah parvaad-vivar-jita-api"//72//

With its import to be such that 'its occurrence is beyond any apprehension as well as it is neither attached or detached to any one' It happens to be same that in present paradigm, it exactly resembles to what we expect of the term 'random numbers'.

Hence Kapil's aspiration having been expressed several millennium years before, of a concept similitude to 'random' appears to have been fulfilled now for the first time through this small and simple text by its Authors. That by the applications of LILA-BHAS trios and other known quadratic Diophantine Equations coupled by the use of Hilbert Space and elementary topological concepts and techniques.

8.8: Non-relevance of Statistical Test for such RRNs to be genuinely random number:

It is to be recalled that most of the statistical tests are unidirectional meaning by rejecting the Null-Hypothesis one accepts its alternative hypothesis, which is equally prohibitive by Kapil's verbatim. Also, in case it is drawn or derived from unintended sources it is bound to have maverick nature. That does not need any further tests. A real RANDOM is indeed a RANDOM. Even though, with smaller of 'n' the observations/ measurements, its scatter may not look so.

8.9: Concluding Remarks:

Thus the contents of this simple and slim booklet are not only its innovative findings, but also of putting GIANTS inside NATURE'S BOTTLE.

Yes, the 'Nature' in its Features conceals largest density of Real Random Numbers (RRNs).

Impact of such results needs to be high-lighted in Chapter 10, that of Epilogue.

References

1. Debnath, L. and Mikusinski, P.:(2006): Introduction to Hilbert Spaces, 3rd Edn., Academic Press, Elsevier India Pvt. Ltd. New Delhi, India

2. Simmons, G.F.:(1963): Introduction to Topology and Modern Analysis, McGraw Hill Inc., New York, US

3. Tripathi, R.S.(1982): Maharishi Kapil's Sankhya-karika of Iswar Krisna: Chowkhamba Krishnadas Academy, Varanasi, UP, India

CHAPTER 9

Nature, Life and Living

9.1: The mysterious ternary: Nature, Life and Living:

Such a tri-variant ternary is so mysterious , yet so rampant that all literatures written so far, from creation to present- time have been unable unravel them. One of their aspects happens to be characterized as 'random' often to be of use by modern scientist as recourse to their ignorance, to face their encumbrance being, the robust decision.

A sentence of Ode to the same:

'O, Nature! Indeed you are fathom-less, vast and varied. So far even your-self remain so encrypted that almost everyone needing you, have the compulsion to use Pseudo Random Numbers (PRNs).'

Earliest this was realized by Maharishi Kapil in his 'Sankhyakarika-72'as herein below:

"Akhya-yi-ka virahiaitah parvad-vi-varjita-chapi |72|"

Its import: It is to be beyond any anticipation and equally unattached from any one, which is nearer to it or away from it. It impliedly hates none also loves none. Though with, in sequence with them, but remains unconcerned with any of them.

Such a concept persisted in Kapil's Sankhya philosophy which is quite similitude/congruent to the concept of RANDOM. The same is the dire need to fetch unbiased decision /selection to encrypted communication from a source to sink. Yes, we have repeated this sermon perhaps more than twice. That is because its relevance could be realized and its importance could be cognized in Indian philosophy much earlier. That is why we are required high light its importance, in the era of Artificial Intelligence, to make man kinds decision robust, beyond the boundary of time and space..

Yagnavalak who enunciated to king Janaka and his learned wives Gargi and Maitri as also with others through Brihad-aranyaka Upanishad as to how wonderfully the NATURE combines or synthesizes several features and factors to form this World and its surroundings, to operate this cosmos. Amongst them were Brahma and Sarasvati who jointly created the Universe with its living beings.

Thus the 'nature' and 'random' are seen to form a lively combination as naturally as though they are 'cohort' of each other in 'cosmos'. In Vedic literature for the term EARTH, its translate is not Prithawi but Prithawi-adyo implying the EARTH and its CONTENTS and SURROUNDINGS. I am the senior author who have a version of Upanisdas translated by Max Muller accredited the same to Sultan Dara Shikoh, who translated it to Persian language.

It is there the word 'aranyaka' has been given the connotation of forest. Otherwise also the word 'aranya', is taken to mean forest or wild to imply 'nature'. As its virtues are 'wild' for the same reason, its adorned name is 'random'. Though one connotes it to jungle: but its presence is noted even in human behaviour, market trends, supply and demand, occurrence of epidemics, earthquake, cloud bursting to almost every non-deterministic events, features of societies, activities not coming under fully controlled decision and/ or activities, those being rarest of the rare . However we may begin with natural entities, events and/ or scenarios.

The Nature, whose landscape picture is attached at (Appendix 1). It is often known as wild, has to have infinite range of genus, species and phylum with wider range of observable and/ or measurable features. Plant height, creeper length, number of branches and shoots leaf-length and/or breadth in terms of digits and/or numbers. Twigs and branches from where to measure, may be chosen arbitrarily, haphazardly or even by RRNs and/ or PRNs (Pseudo Random Numbers) already available. What is called 'wild' for its aspects when considered as of use for robust decision making gets the connotation of 'random' as expressed earlier. Also when any one is in hornet nest (the dilemma, two horns of syllogism well defined in deductive logic as many of us take recourse to tossing a coin) and unable to take decision he virtually surrenders to 'random'. There is reciprocal call as though of 'random', assuring come what may: 'I am as nature, always with you'. Actually it (the random) gives at least a good start and pathway to proceed ahead. Therefore it was decided to proceed with actual observational experiment to draw directly RRNs out of features of nature.

9.2: Direct drawing of RRNs from features of natural entities:

To show how RRNs can directly be derived from measurements of natural entities.

Example 1:

On our front door, we have Money- creepers (Epipremnum) having thousands of twigs and its leaves of different lengths and breadths, each of them being measurable in centimetre and millimetre. There from we selected with our eyes- shut from here and there haphazardly but not using any random table, to pluck them. There by measured their length and breadth and obtained their product in order to get minimal- area rectangles to contain each of them separately. They were for 25 leaves. They are tabulated as below in Table 9.2.1., in order of their selection to pluck them.

Table showing Length and Breadth in Cm . mm and of their product as Areas in Sq. Cm,s. Thus the real random numbers (RRNs) are being naturally fabricated as, RRNs = Areas×100.

TABLE 9.2.1

Sl. No.	Length	Breadth	Product as Area	RRNs
1	17.1	14.8	253.08	25308
2	10.8	8.9	96.12	9612
3	10.9	7.5	81.75	8175
4	9.0	10.0	90.00	9000
5	13.1	12.2	188.82	15982
6	16.9	15.8	267.02	26702
7	11.2	8.8	98.56	9856
8	12.5	10.4	130.00	13000
9	12.3	11.4	140.22	14022
10	12.6	10.5	132.30	13230
11	21.1	20.0	422.00	42200
12	7.0	7.8	54.60	5460
13	10.5	7.0	73.50	7350
14	42.1	36.5	1536.65	153665
15	8.5	7.5	63.75	6375
16	9.5	10.4	98.80	9880
17	9.9	9.0	89.10	8910
18	10.8	7.4	79.92	7992
19	39.0	33.2	1294.80	129480
20	19.0	18.0	342.00	34200
21	15.0	13.0	195.00	19500
22	18.3	16.8	307.44	30744
23	35.9	32.0	1148.80	114880
24	25.7	22.8	585.96	58596
25	9.9	9.0	89.10	8910

RANDOM SCATTER

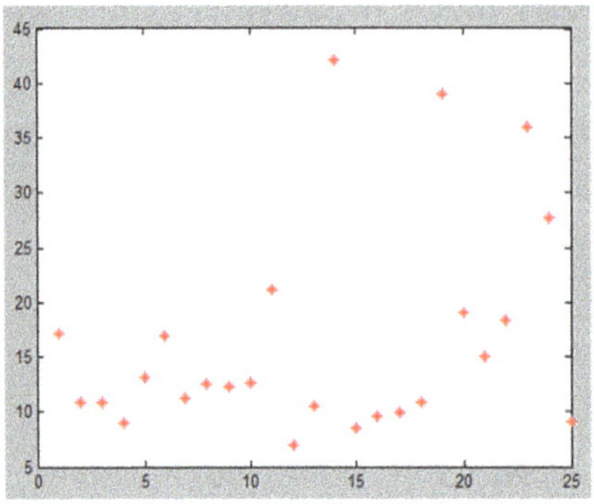

RANDOM MAGNITUDE

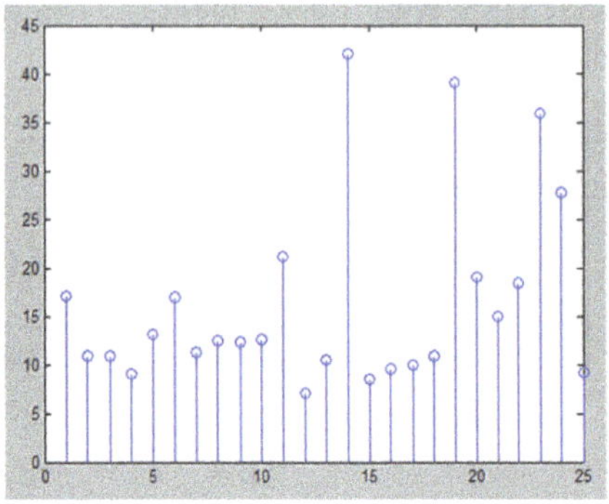

Example 2: Another example is on betel- leaf (Piper betel):

We have betel-leaf plant that has been grown in earthen- pot. Though its leaves are smaller but has copious twigs and leaves. Here we have taken length of randomly chosen 20 leaves but independently of them we have breadth of 21 leaves, both in cm and mms. They are presented tabular format in Table No, 9.4 and 9.5 respectively.

Table No. 9.3

Sl. No.	Leaf-length	Random No.
1	3.8	3800
2	5.8	5800
3	4.2	4200
4	3.5	3500
5	4.7	4700
6	4.6	4600
7	4.1	4100
8	3.9	3900
9	3.5	3500
10	4.8	4800
11	5.5	5500
12	5.1	5100
13	6.1	6100
14	5.8	5800
15	4.0	4000
16	3.9	3900

17	5.0	5000
18	3.7	3700
19	5.3	5300
20	6.3	6300

RANDOM SCATTER

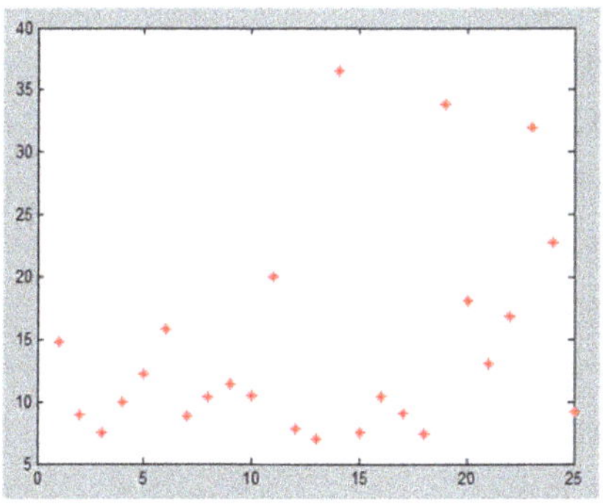

RANDOM MAGNITUDE

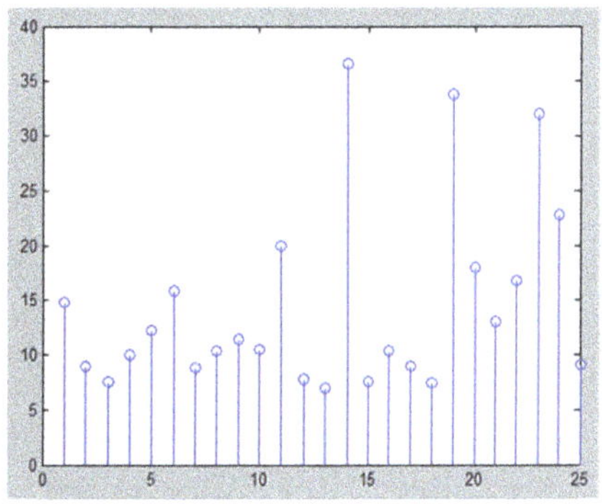

9.3: Examination of observation/ measurements:

Let us now examine whether, Column 5 which are only 100 times multiples of Column 4 appear to contain RRNs. Those have been derived directly by measurement of natural entities.

We are required to assume that every such natural entities are naturally random and those do not need filtering. This, what we assert, that random is random, nothing else. Those are RRNs.

RANDOM SCATTER

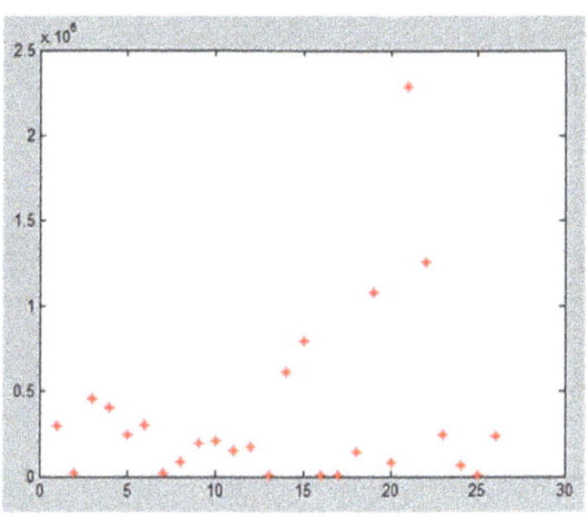

RANDOM MAGNITUDE

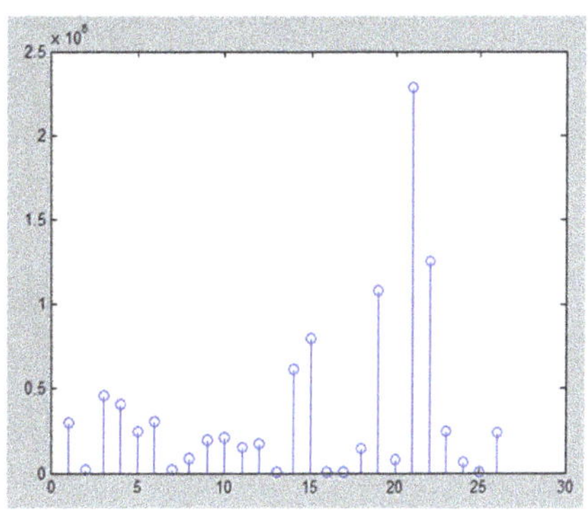

Example 3: We take another example on total river-flow data derived from mean flow data from Rao and Hamed (2000). They were obtained larger recording points (RPs) but we have taken for 1 to 26, RPs only.

Table No. 9.4

RPs No.	Total flows
1	153069
2	144900
3	244729
4	170064
5	78048
6	245718
7	4558774
8	1252696
9	2232
10	193995
11	15228
12	795532
13	301400
14	107968
15	59363
16	296276
17	5607
18	405622
19	280992
20	5842
21	240590
22	17052
23	4246
24	207180
25	607922
26	86730

RANDOM SCATTER

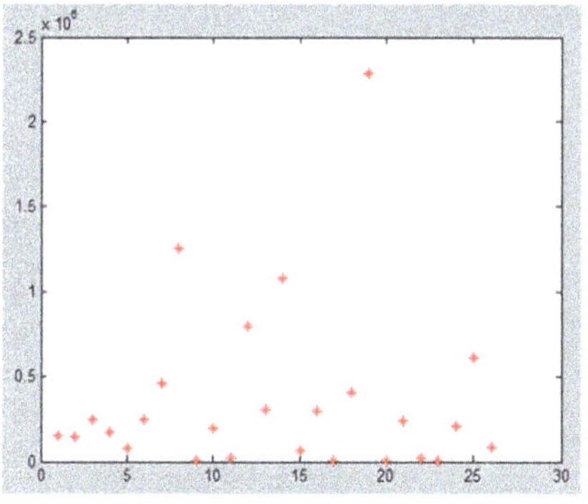

RANDOM MAGNITUDE

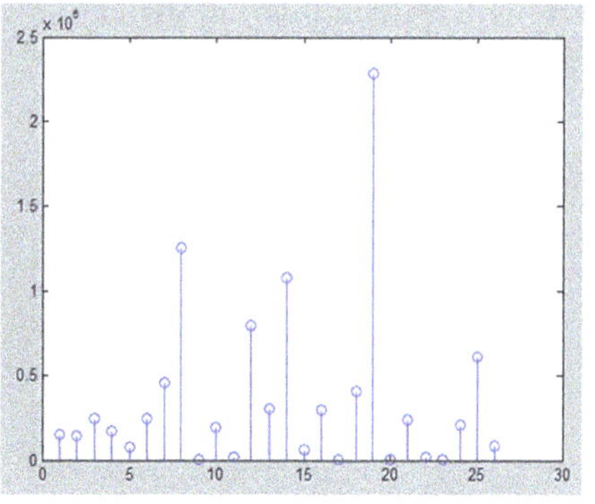

The annual total flows of water are it selves taken as random numbers (RRNs) to be selected so for any purpose akin to problems to which generated pertains. The Graphics illustrated may indicate the nature of such random for their scatters and magnitudes.

As the nature is vast and varied, it is capable of containing and displaying much larger numbers than large prime numbers, at the same time not needing Modular Arithmetic. Also able to break philosophical paradox persisting at least a century back. It is not like killing three birds by throwing single stone chip: but simple, short and cheaper device to provide viable alternative to learned peers' several existing devices. Those include millions of research publications and billions/ trillions of dollars, pounds and/or Rupees and even other monetary units.

9.4: Inference drawn from observation/ measurements:

That indeed opens vast and varied field for getting RRNs. But often, too good is too bad, as the same raises problem of choice, requiring criterion for choice. It is indeed difficult. Only clue that can be hyped is to know that RRNs, those are appropriate for occasion are very near your problem as it must have the same scale to which your current data belong. That further takes you away from PRNs but closure to RRNs as you are to derive the same from residue of your observation or measurement.

9.5: Attributes/ characteristics required in entities being observed:

One more pertinent question that may arises, is that RRNs must be quantifiable i.e. to be expressible in digits or numbers. Incidentally that has been solved much before as non-quantifiable attributes may be put as strings and ultimately may be digitized as 0 and 1.

Tables and graphs do indicate authenticity of their being RRNs but far from being equi-probable as its coming in sample is nearly in proportion with its density of occurrence of the population wherefrom sample entities are drawn. That could have happened, had entities been rectangularity distributed. Thus RRNs so obtained could be very useful for encryption but lesser for drawing finite sample to represent population unless the sample choice could be probability proportional to its densities of occurrence. Statistical Sampling Theory, do very well consider to draw sample that way also.

However we have alternative recourse to choose the courses of 'approximation' and/ or of 'filtering', presented in Chapters 5 and 6, and also those that are presented though Chapters 7 and 8.

9.6: Scope for its applications:

Let us recall that in present paradigm shift, where the term random or its other formats, may find scope for application, just to gauze how wide is that of our innovative device. They are better indexed by Knuth(2000) in has Classic Work, as here in :

1) Random Bits 2), Random Choice, 3) Random Combinations, 4) Random Deviates, 5) Random Devices Directions, 6) Random Fractions, 7) Random Functions, 8) Random Integers, 9) Random Mappings, 10) Random Numbers, 11) Random Number Generators, 12) Random Processes 13) Random Real Numbers, 14) Random Variables, 15) Random Sample, 16) Randomization and 17) Randomized Algorithm, Including 18) Random Clusters. May be some more like coding to cryptography, where it provides philosophically sound base for their ever rising citadel. As aspired at its Preface of this short simple monographic text.

Thus our innovative device may find its scope in all those to change their features or patterns from PRNs to RRNs. Indeed a commendable development, for its economic, simple and time saving aspects, presently a 'transition'. But no sooner it is able to minimize or replace PRN, it can claim to be an 'evolution'. It might never be a 'revolution', as the same has sense of violence.

9.7: The Nature within:

Nature has to have non-enumerable features inside like plant, animates including human bodies. They are like body-temperature, blood-pressure, blood-sugar, bleeding-time, clotting-time, vitamin and mineral contents, its secretions and discharges etc. Like those there are many other aspects and features where comparative data are made available for normal individual verses abnormal ones. A good deal of randomness, are seen to exist in such data-sets also. In absence of data it would not be possible demonstrate such estimation processes as have been presented above. Nevertheless, approximation algorithm (herein Chapter 6.), on normal vs. abnormal may possibly be made applicable here also to decipher their differences as random numbers.

9.8: The Nature outside:

In present paradigm, shift nature includes, market demands and supplies, input and output , birth and death, rich and poor, manufacturing and marketing data and say export and import etc. . So to say inexhaustible such data – sets. The term 'random number' may possibly be required everywhere. Thus our effort about being made here, how so over large, may not be even a drop in ocean.

9.9: The Tyuchev's expressive lines:

To authenticate our feeling towards such Nature's Random, we quote Tyuchev's lines which as motto of Wolkenstein's (1969), Holzmuller's (1981) and now our own's purpose, it the Nature.

"Nature is not what you think,

it is not a cast, soulless face.

It has soul , in its freedom,

in its love, it speaks to us."

The author adds here ," What it likes and/or dislikes."

9.10: Much more examples:

We need not have more examples as Nature is so vast and varied that any one may choose sample of his choice to count and then measure its features and get random number akin to his/her problems. All those will be RRNs not PRNs.

9.11: Natures' vegetative photograph:

We place here at Appendix 1, photographs taken from our house of the picturesque-scene of how forest-green combines blue-sky, making random vertical lines, beautiful Real Random Numbers (RRNs). But rugged random sequence of vertical lines of green forest, are all independent as none of such sequential entities affect or effect, one another.

9.12: Being on appropriate for RRNs for ones need:

How and where from to search for the RRNs of one's need for the occasion?

Since you have the data-set on your problem in hand but you require RRNs, matching the same. The golden rule is to choose the remains, remainders, remnants, residues, rubbish, rudiments that you are required to ignore while approximating and/or filtering, the data-set . To summarise it is to be said that random number one needs is very near his problem, it just a throw- away distance from targeted ones.

Again a realistic simple clue is to click around your problem for which such random number is being sought.

Besides the Authors here have large sets of data both for

1) Approximation

   And

2) Filtering,

Algorithms/device obtained, not related to problems of any ones requirements, as they have been obtained much earlier in course of his research needing iterative searches. From the same RRNs are to be derived on any specific requirements.

Though RRNs obtained from nature's features are truly random but not equi-probable random, instead it happens to be randomly random. It is so not because of its fault but of the nature, as its features are rarely equi-probably distributed. Thus remains abruptly useful for encryption in any of its five formats contained in this monographic text.

References

1. Holzmuller, W.(1981): Information in biological systems: the role of macromolecules, Cambridge Univ. Press, Cambridge, London, UK
2. Knuth, D. E.(1998,2000): The Art of Computer Programming, Vol. 2, Semi-numerical Algorithms , 3rd. Edn., Addison- Wesley (Pearson Education), Delhi, India
3. Rao, A.R. and Hamed, H.H.:(2000): Flood Frequency Analysis, CRC Press, Boca Raton, London, New York, Washington, D.C., US
4. Wolkenstein, M.W.(1969): Molekule und Leben. Leipzig: VEB Ge

CHAPTER 10

Epilouge

10.1: Introduction:

We have made quick and short references for the terms RANDOM, RANDOM NUMBERS and of RANDOMIZATION, as available in our personal library. The topic is so vast and varied that Knuth (2000) took nearly 200 pages of his voluminous text and more than thousands of references about the same. Other texts referred here of numerous learned peers are of no less importance, reflecting the same, besides its inescapable uses and/or applications.

Incidentally, virtually all of them happen to be the users of PRNG (Pseudo Random Number Generation), in spite of well known philosophical paradox, which is preventive, yet persist its uses.

10.2: Incipient for first four chapters:

Authors there after expressing indignation to such persistence but inescapable paradigm in first four chapters, there from have been able to carve out five devices (modus operandi) through which Real Random Numbers (RRNs) could (possibly) be obtained.

Why the term possibly? This is the term every empirical scientists is required to use. Unless any process meets large scale use, one need not be emphatic about any empirical processes being adopted, in spite of the method may have some advantageous features, well demonstrated .

10.3: Incentives to proceed ahead:

RRNs so obtained take due care of almost century old of the said paradox: instead of any objective to obtain them, they make,

1) Use of errors obtained while making approximations, being considered negligible by some remote approximate seekers,

2) Use of residues or remains of filters, considered worthy to be thrown by filter seekers,

3) The same through multiple applications of available Quadratic Diophantine Equations based on Herkommer's (1999) primitive Pythagorean Triples: the first one was discovered by Bhaskarachaya (1151) as published in his book LILAVATI.

4) The same through use of complex (conjugate roots) pairs of Topology as multipliers of 3) herein above, instead of real numbers there. The same, is seen to create distinct Hilbert Space of solutions, those here happen to be much sought after RRNs.

5) By counting, observing and measuring natures widely varying amongst species and/or phylum and between them.

At this juncture a question was raised as to 'why hilbertization, virtually orthogonality is considered necessary?'

The quick answer is 'we are often required to know the independent effect of every effort we make or every cause of change we apprehend'. In its current higher generality of mathematics it is being looked on as canonical isomorphism, as expressed in Representation Theory. Refer Fulton and Harris (2004).

While in usual medicinal practice it is required to be observed that a medicinal doze has to have no side effects. For scenario of warfare it required to be observed that only army arsenal are required to be targeted, not civilian area gets affected.

It is a conceptual dogma not a physical reality, though may be aimed, at but seldom attainable. It is so as all earthly objects are basically synthetics which waits to be analysed to assess the role of every analysed components.

As and when needed some such components are mixed-up called strategic mixture or mix-up to meet some prefixed objectives. Though mostly these features are well known but not the way we have dragged them.

Thus we the authors feel satisfied that we have been able raise dependable generators for RRNs, perhaps by eliminating or at least minimizing century old philosophical paradox. Indeed it is an extra- ordinary feat.

In furtherance of our such feat, we may say now, every one may get RANDOM NUMBERS of VARIABLE DIGITAL LENGTH, in accordance to its changing requirements for CODING, ENCRYPTION as well as for various other purposes.

Hence, here we are proud of optimising the optimum.

At the same time we feel that we placed before the World only those aspects of Nature that are visible but not what could be heard, smelled , tasted and touched. In them also elements of random would, very widely be contained.

Sooner it is widely accepted better it is for ubiquitous use. No doubt it is one of the TIME and SPACE CONQUEST.

10.4: Notable points indicated:

Though, outside the scope of this monographic text, notable point to be indicated is " Data is not mere number but quantified information. The other face of the coin is 'a number is not mere number but a vector to carry information expressed in number, being transform to data.' "

Often such aspect gets missed even by learned dignitaries, when they be non-serious mood.

10.5: Clearance of doubts raised:

One more doubtful question has to be cleared.

That is on 'In cosmos human being is required to face like events, entities, happenings, instances or occurrences , are to be taken as continuous or discrete?' Such ticklish question is being raised since time immemorial. Even Einstein's famous papers from his Annus Mirabilis of 1905 provided the explanation that 'light was actually made up of discrete packets of energy'. Refer, Whyntie & Pugh (2013).

Shannon similarly established through his two theorems that Information is also discernable and be transmitted discretely. Refer, Jelinek (1965) or many other texts.

In existing computer era almost every- one is required to know that all features needed to be inputted to it are transformed to 0 and/or 1, thus discrete.

Thus every phenomenon of this earth has to be considered as discrete and nothing of them may be continues.

But phenomenon of continuity verily exists.

As example it is to be said that our planet the earth is spinning continuously on its axis and that in every twenty-four hours it takes a complete turn. Thus a point on it near its equator travels more than 67,720 km a day by way of its spin.

At the same time it is moving round the Sun in every 365.25 days: a distance much longer than above distance round the Sun, called Aphelion of travelling distance 152 million Km. Refer, Jones (1984) and Jazwinski (1970 & 2007).

Those two such speedy movements perhaps do not cause any jerk to be felt by its inhabitants. Therefore could be regarded as continuous phenomenon. Further it has invariably continuously constant property of gravitation. At the same time its contents: the nature, with all its living beings and/or identifiable organic substances, are both discrete and stochastic. Hence the 'random number' has to acquire the same features as its various contents.

10.6: Our last sentence for the book is to say that 'you may start writing a book but it not easy to finish the same'. You have to cut-short, at a point, to finish the same.Where the author feels like cut-throat.

10.7: Also, it is here only that the author would know how and where he should have begun the writings of the book.

References

1. Bhaskaracharya :(1151):LILAVATI, PATIGANITA the Mathematics, Edt. by Dr.Rama Shanker Tripathi (1982), Chowkhamba Krishnadas Academy Varanasi, UP, India

2. Fulton, W. and Harris, J.(2004): Representation Theory: A First Course, Springer International Edition. New York, Berlin, New Delhi .

3. Herkommer, M. A.: (1999): Number Theory: A Programmer's Guide, McGraw-Hill, New York, New Delhi, India

4. Jazwinski. A.H.:(1970 &2007): Stochastic Processes and Filtering Theory, Academic Press .Inc. & subsequently by the Dover Publication Inc, New York

5. Jelinek, F.(1969):Probabilistic Information Theory, Discrete and Memory- less Models, McGraw-Hill Book Company, New York, New Delhi

6. Jones, B.W.(1984): The Solar Systems, Pergamon Press , Oxford, New York, Toronto, Sydney, Paris, Frankfurt

7. Knuth, D.: (2000) : The Art of Computer Programming ,Vol.2, 3rd. Edn. Addition-Wesley , Delhi, India

8. Whyntie, T. and Pugh, O.:(2013): Particle Physics a Graphic Guide, Published by Icon Books Ltd., 39-41 North Road, London N7 9DP, UK

APPENDIX 1

Random From Nature

When we think of NATURE, the first thought that comes to my memory happens to be FOREST. It is that which has been herein below its photographic- panoramic view. The same was ushered in literature through **'Brihada- aranka Upanisad'**, where Rishi Yagyavakya explained to his two wives Gargi and Maitri and others, the secrets of the NATURE. Those which have been, the creation of Brahma and Saraswati. Their stories about NATURE happen to be endless episode, . To know about the same reader is advised to refer the said Upanishad.

Let us refer to modern thinkers about the same. Of them we choose one the Wolkenstein (1969), who exposed his thought about the NATURE:

"Nature is not what you think, it is not a cast a soul-less face. It has soul, in its freedom, in its love, it speaks to us."

Let us conceive as to what it speaks about RANDOM. It directs us to have a look at the picture here in below.

Yes, just behind my house the trees are in SEQUENCE amongst them, only any- one is neither to be regarded as FOREST nor RANDOM. It is only their existence with others in SEQUENCE make them both the FOREST and RANDOM.

However, none of them is dependent on others nor they have any regards or disregards towards any of them. They grow, they exist, they may decay, they shed their leaves, they yield fruits for none of them: much less their seeds do or don't give birth to any other of them.

That are features on what they are considered any one of them being called orthogonal to rest of them. The same in to-days time and space sequence are called members of Hilbert Space. For the same Maharishi Kapil had the hype for his concept, congruence/ similitude to/of RANDOM as quoted several times by us in Chapters of our this very text. Also for the same, Bhaskaracharya in his LILAVATI,(a well known text on Indian Mathematics) contained the orthogonal trios 3,4 and 5 of digits: which imply $3^2 + 4^2 = 5^2$. No other digit-trios could display such feature except their multiples.

Yes, not to be forgettable Pythagorean property of RIGHT ANGLE TRAINGLE was proved to be true to establish the property of orthogonality, geometrically.

We hope we have been able to make our readers in few words as to what the term TRUE-RANDOM means and how Real RANDOM NUMBERS (RRNs) may possibly be obtained from FEATURES of NATURE.

APPENDIX 2

On Discrete Normal Distribution

The graphic representation for the frequency distribution is presented in figure here in below.

It represents discrete prototype of Normal (Gaussian) Probability Distribution. Since in real world situation most features that are being required to face are discrete, we may very well replace this in place of traditionally being used the Normal Probability Distribution which has been for continues variable.

This newly found distribution ,apart from being symmetric is finitely terminating. Thus now no one is required to assume for sampling distribution that the sample has been drawn large population following Normal distribution. Of course the variables now required to be standardized to the mean 0 to lie between (-1 to +1) and the variables may assume values: -1, -0.9 ,-0.8, -0.7,-0.6, -0.5, 0, .5,.6,.7,.8,.9 and 1.0 with sum of their relative frequencies be equal to 1.

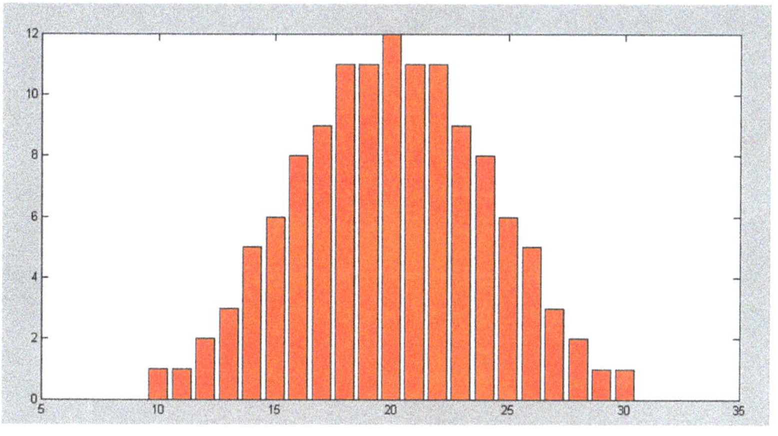

Figure 2.1

Here the horizontal axis X stands for Vector Sums and the vertical axis Y for the corresponding frequencies of occurrence.

Such ***Yantrik*** frequency distribution is perhaps the first prototype/ similitude of Gauss's Normal Probability Distribution, which has just emerged almost unexpectedly/ incidentally It is to be examined from comparative features of Gaussian (Normal) probability distribution.

- Apart of its symmetry, its pick point is at the middle of its finite terminal points at both ends.
- Its most notable feature is, its asymptotic look at both ends with non-changing frequencies being 1 and 1 the symptom of asymptotes at both end. Also like Gaussian Standard Normal, its, mean=median=mode = 0.

To summarize, its distinguishing features are its finite termination. At the same time, above all it is revealed that, without any assumptions of the variable being continuous, population being large to reach infinity asymptotically, it exists even without requiring the famous the CENTRAL LIMIT THEOREM. Further without using π and é or to approximate factorial of any number by Stirling's expression.

Here appears that of its standardized form,(i.e. zero mean and unit variance) the Data Scientists of today may like to use the same in empirical inference and testing of hypothesis, as alternative to tests based on normal probability distribution.

Subsequently we shall take the same for data based examples. In case, test based on this distribution is taken up, we may neither have to do continuity corrections nor transformation of data for its adherence to the normal distribution. However level of significance have to change from traditional 5% and 1% to 8/126 = 6.3%, 4/126 = 3.15% and 2/126 = 1.6% approximately for two sided tests due to discrete structure of the distribution.

Such a frequency/ probability distribution is fairly good prototype/similitude of the Gauss's Normal Probability Distribution and best suited for discrete variables, which is proving grim reality in present paradigm shift.

Its standard transform is:

M = | x – m(20)| / sd(4.099), where m= 20 and sd = 4.099 for given/obtained data- set.

The test based on M may be a discrete version of Z–test , the usual normal test, with its level of significance changed as per suggestions made in preceding paragraph.

The Author was astoundingly surprised to get prototype/similitude of Gaussian distribution which may have far reaching paradigm shift on existing Statistical Inference.

Importance of Yantrik Distribution

The continuity of variables and of infinity, are in abundant use in Mathematics, and in Biological Sciences, Engineering / Physics/Statistics and in their applications. Equally required are of discrete features claiming to be the grim realities of life , livings and many other aspects of nature where either continuity or of assuming as large as the same tends to infinity has no or little realistic relevance, are not required to be discussed or opened here. That may be done in separate publications. This is also because the scope of the booklet is limited to:

1) Find some designs for decision for start up guild, fulfilling M-optimality criterion that may involve limited coalesced cum independent ventures at the same time provides hedge to risk.

2) While in making attempt to achieve such multi-objectivity of the designs for decisions, the use of Markandey and his type of Yantras of being made.

3) Also to establish linking of such designs to "Representation Theory" for being treated as group acting on its connected vector spaces, Lie algebraic features, geometry and topology as well as their plethysm (implying its multiple existence) without change in their desired characteristics.

4) Finally to validate such designs through data-based examples (without needing any proof), adhering only to step-wise algorithms.

But both of them that continuity and infinity are beyond realities of life, livings and materials to be of use: whether their forms are solid, liquid, fluid, molecules, atoms, and pixcels and more so of bits and bytes. As the latter does not exists, while the former is phenomenon perceivable for moving objects where it is to be regarded to realities. Therefore and all that can be expressed through them, are of discrete structures. Even fractals and wavelets illustrate the same. They have definite size, but while acting in their large numbers, Einstein(1905) and his graduate assistant Jean Perrin (1913), found that such tiny cells suspended in well mixed emulsions, acting in large numbers follow Brownian(1827) motion. Those were seen to follow Gaussian (1809) probability distribution, there by the necessity for existence or of search of any of its discrete similitude could easily be escaped through. But it is to be noted that their count of suspended particles was as large as 17,000 and so large might be of motion pictures and of T.V's reflections. Reader may refer to MacKinnon's paper(2005).Who refers to Einstein's 1905 paper for his two decisive innovations.

"The first was to extend statistical methods to fluids.

The second, a manifestation of unique creative insight, was to ferret out a problem where assumption concerning the size of molecules would lead to observable consequences.

Einstein isolated a problem, tiny particles suspended in fluid, that could be treated in two different ways. The first involved treating the fluid as a collection of molecules rather than a continuum. (clearly a discrete phenomenon)."

Thus who may desire freedom from the yoke such assumptions, as are required for normality of distribution, the Yantrik- distribution found here in preceding Sections 2.4 and 2.5, offers the realistic recourse. Authors here both Brown and Perrin could have counted such particles before and after dilution or prepared emulsions to ascertain the conclusive findings along with references that are available in an article by Mackinnon on "Einstien's 1905 Brownian Motion Paper" CSI Communication Dec 2005.

Incidental discovery of Discrete Prototype/Similitude of Normal (Gaussian) Distribution: its importance. More about it are obtainable from Das (2023).

Reference

1. Das, N.C. (2023): Entrepreneurial Designs for Start-ups, Shorff Publishers & Distributors Pvt. Ltd. Navi Mumbai , Banglore, Kolkata and New Delhi

APPENDIX
3

Maharshi Kapil's Shasti-Tantra

1: Maharishi Kapil in his 72nd , couplet of his Sankhya-Karika, while explaining his concepts being congruence/similitude to RANDOM, several millennium years before he wrote:

सप्त्यङ् किल ये अर्थास्तेऽर्थाः, कृतसंन्यास्य षष्टितन्त्राः।

आख्यायिका बृहद्‌हिता परवद्विवर्जिता तास्य चापि॥ ७२॥

Saptyang kile ye-artha-ste-arthah kritasna-shya,shashti-tantraah

Aakhyayika vrihada-hitaa parvad vivarjita-tasya-chaapi |72|

Before explaining his thoughts congruence/similitude to random, refers to his concept of Sashti-tantra , that being underlying concept towards Sankya philosophy, drew the attention toward the same. Those were his triplets of Gunas the virtues. Those having been Sato-guna, Rajo-guna and Tamo-guna.To the same every human being are required to pass necessarily some time in his|her life span. But those occur haphazardly, hence congruent to RANDOM.

Guna	*Phases*		
Virtues	*Weak(-)*	*Strong(+)*	
1: Sato-guna	-	+	*The truthful stage*
2: Rajo-guna	-	+	*The enjoyable life stage*
3: Tamo-guna	+	-	*No painful life span in a phase*

Phase

This makes $3 \times 2 = 6$, the six- stage system, (sasti-tantra),the hexagonal systemic phases. It could be different also in one's life span, like opposite in signs (+) and (-).Those could be torturous to pass through.

Only on such life span a saintly person may acquire the personality, of being able act or face any aspect which is congruent/similitude to random: who remains beyond any aspiration and unattached to any aspect of life or object he is required to face.

It is beyond the scope of this text , however when concept of congruency/ similitude to RANDOM had to be faced, we had have to interpret the same to our limited knowledge. There could be some more ways to interpret it. Those have been explained in Bhagvat Gita , in Durga-sapt-sati and/or by many other great philosophers.

References

1. Tripathi, Dr. R.S.: (2007): Rishi KAPIL'S ' SANKHYA KARIKA' Chaukhamba Krishnadas Academy Varanasi, UP, India

APPENDIX 4

Comparison of Scatters for PRNs to that of RRNs

The Appendix aims to present comparative pictures of the two conceptual entities of RANDOM NUMBERS. The first one that of PRNs and the second for RRNs. For the first we obtained random numbers from Kendall and Babington-smith (1939) Table. From the same we drew randomly only hundred numbers from its the very first thousands random numbers. Thereafter we selected only thirty serial sequence such numbers to be made comparable to our very first sample obtained in Chapter 5, by approximation mechanism.

TABLE OF (PSUEDO) RANDOM NUMBERS: (1 TO 100)

	1	2	3	4	5	6	7	8	9	10
1	23	15	75	48	50	1	83	72	59	93
2	76	24	97	8	86	95	23	3	7	44
3	5	54	55	10	43	10	53	74	35	8
4	90	61	18	37	44	10	96	22	13	43
5	14	87	16	3	50	32	40	43	62	23
6	50	5	10	3	22	11	54	38	8	34
7	38	97	67	49	51	94	5	17	58	53
8	78	80	59	1	94	32	42	87	16	95
9	97	31	26	17	18	99	75	53	8	70
10	94	25	12	58	41	54	88	21	5	13

Now have a look at the SCATTERS and its corresponding MAGNITUDES yielded by a sample of PSEUDO RANDOM NUMBERS which follows those of REAL RANDOM NUMBERS obtained in Chapter 5 as well as FRONT COVER DESGN also. They both are of the same numeric size (30) to make them comparable for their (SCATTERS) or HAPHAZARDNESS a term used by Knuth to gauze the degree of RANDOMNESS along respective MAGNITUDES as stated also herein above.

RANDOM (PRNs)

SCATTERS

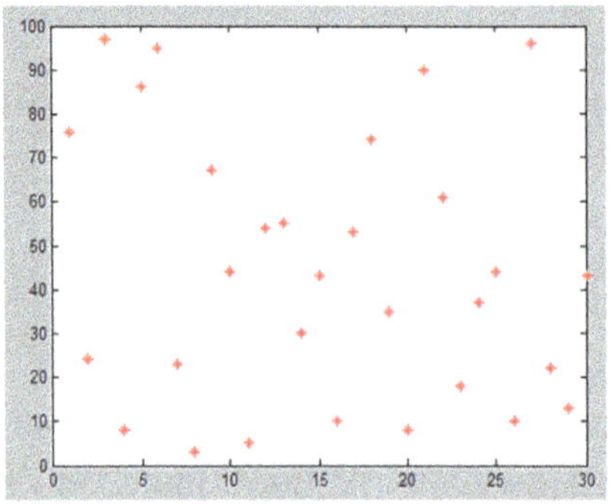

MAGNITUDES

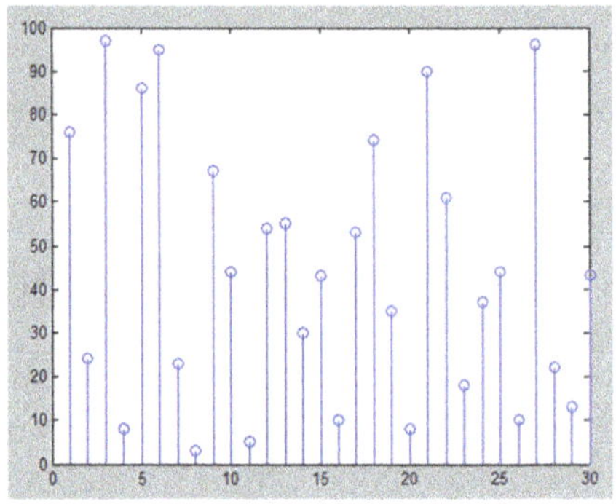

RANDOM (RRNs)

SCATTERS

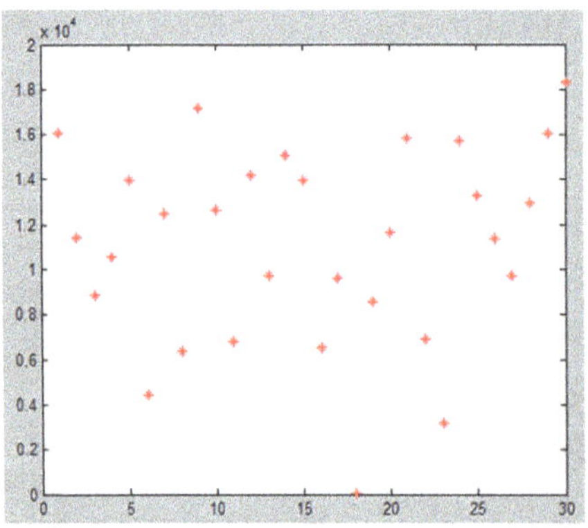

MAGNITUDES

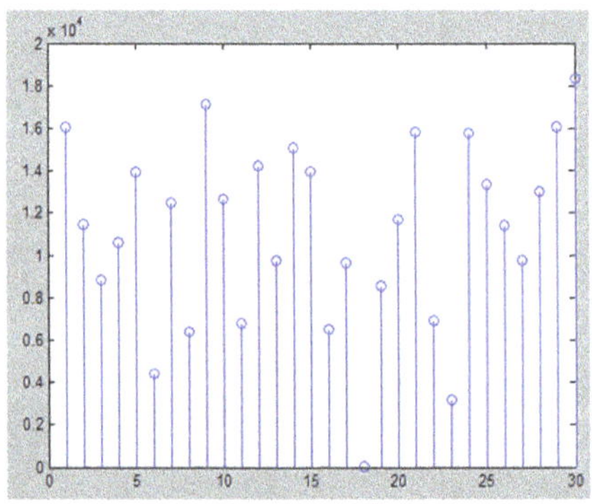

The purported comparison clearly asserts that the latter(RRNs) in no way inferior to former (PRNs) which are being regarded as EQUALLY PROBABLE RANDOM numbers or to have emanated from RECTANGULARLY DISTRIBUTED POPULATION. By implication it is to be inferred that even RRNs may be regarded worthy of the same virtue but on the other hand completely bereft of its century old vices of philosophical paradox. Of course such hype is based on empirical finding. Mankind have to be satisfied with empirical findings as absolute finding are mostly Utopian the concept being mere imaginary.

INDEX

O

P

Q

R

S

T

U

V

W

Y

www.ingramcontent.com/pod-product-compliance
Lightning Source LLC
Chambersburg PA
CBHW061356240926
51846CB00007B/1424
9789368080930